DEADLY CHOICES

How the Anti-Vaccine
Movement
Threatens Us All

PAUL A. OFFIT, M.D.

BASIC BOOKS
A Member of the Perseus Books Group
New York

Designed by Brent Wilcox

The Library of Congress has catalogued the hardcover as follows:
Offit, Paul A.
    Deadly choices : how the anti-vaccine movement threatens us all /
Paul A. Offit.
        p. ; cm.
    Includes bibliographical references and index.
    ISBN 978-0-465-02149-9 (alk. paper)
    1. Vaccination of children.   2. Vaccination of children—History.
3. Vaccination of children—Complications.   4. Vaccines—Health
aspects.   I. Title.
    [DNLM: 1. Attitude to Health—Popular Works.   2. Vaccination—
trends—Popular Works.   3. History, 20th Century—Popular Works.
W 85 O32d 2010]
    RJ240.O38 2010
    614.4'7083—dc22

                                                        2010022446

ISBN: 978-0-465-02962-4 (paperback)
ISBN: 978-0-465-02356-1 (e-book)

10 9 8 7 6 5 4 3 2 1

PRAISE FOR *DEADLY CHOICES*

"An eloquent and bracing book."
—*City Journal*

"This book is a much-needed inoculation to help prevent future public scares."
—Tevi Troy, *The Weekly Standard*

"[A] convincing exposé of the anti-vaccine movement."
—Rachel K. Sobel, *Philadelphia Inquirer*

"*Deadly Choices* is a book that should be read by any parent worried about vaccine safety."
—*Galveston Daily News*

"[*Deadly Choices*] is a well written, meticulously documented book that calls out the misguided and dangerous activists who are attempting to hijack an important part of our public health system."
—*Tucson Citizen*

"If more people with an interest in knowing the truth read this book . . . we may reduce the mayhem caused by the anti-vaccine movement."
—*Health Care News*

"Every doctor's office should have a copy of Offit's book, giving parents the other side of the story."
—*Science News*

"If you want a solid grasp of the worries, fears, misunderstandings, and ideology that have inspired a small minority of people to vocally oppose vaccination for more than 100 years, Offit's is the book to read."
—*Health Affairs*

"This passionate book is a candid and educational must-read for all health-care professionals, public health officials and especially for parents."
—*The Guardian*, Punctuated Equilibrium blog (UK)

"A smart, hard-hitting exposé of vaccine pseudoscience. . . . Worried parents, especially, will find this a lucid, compelling riposte to antivaccine fear-mongering."
—*Publishers Weekly* (starred)

"Offit pulls no punches."
—*Booklist* (starred)

"Offit takes aim at the anti-vaccine movement in America and scores a bull's-eye."
—*Kirkus Reviews* (starred)

"[This] is a must-read cautionary tale of the abuse of celebrity, the media, misinformation and fear."
—Steve Novella, Assistant Professor and Director of General Neurology, Yale University School of Medicine

"A passionate, but also compassionate, call for civil discourse and rational conversation."
—Perri Klass, Professor of Journalism and Pediatrics, New York University and author of *Treatment Kind and Fair*

"I recommend *Deadly Choices* by Dr. Paul Offit—a timely and courageous call to arms by the nation's foremost expert on pediatric infectious disease."
—David Oshinsky, winner of the Pulitzer Prize for *Polio: An American Story*, and Jack S. Blanton Chair in History at the University of Texas

"If you care about the health of your children—or the health of anybody's children—you need to read this book as soon as you can."
—Michael Specter, Staff Writer, *The New Yorker* and author of *Denialism*

"[*Deadly Choices*] is a courageous book by a courageous researcher and physician."
—Robert M. Goldberg, Ph.D., Vice President, Center for Medicine in the Public Interest and author of *Tabloid Medicine: How the Internet is Being Used to Hijack Medical Science for Fear and Profit*

To Maurice Hilleman and Stanley Plotkin,
who taught me the beauty of reason
and the power of vaccines

The judgment of history is without pity.
—RAYMOND ARON

CONTENTS

PROLOGUE ix

INTRODUCTION xi

1 The Birth of Fear 1

2 This England 13

3 A Crude Brew 25

4 Roulette Redux 45

5 Make the Angels Weep 57

6 Justice 85

7 Past Is Prologue 105

8 Tragedy of the Commons 127

9 The Mean Season 149

10 Dr. Bob 171

11 Trust 191

EPILOGUE 207

NOTES 215

SELECTED BIBLIOGRAPHY 255

ACKNOWLEDGMENTS 257

INDEX 259

PROLOGUE

There's a war going on out there—a quiet, deadly war.

On one side are parents. Every week they're bombarded with stories about the dangers of vaccines. They hear that babies get too many vaccines, overwhelming their immune systems—then they watch them get as many as thirty-five shots in a span of only a few years and sometimes five at one time. They hear that vaccines cause chronic diseases. And they hear this from people they trust: celebrities like Oprah Winfrey, Larry King, Bill Maher, Don Imus, Jenny McCarthy, and Jim Carrey; elected representatives like Carolyn Maloney, Chris Smith, Dave Weldon, and Dan Burton; television correspondents like Sharyl Attkisson of *CBS Evening News*; and popular doctors like Mehmet Oz and Robert Sears. But mostly they hear it from parents like themselves—parents who claim their children were fine one minute, got a vaccine, then weren't fine anymore. Understandably, some parents are backing away from vaccines; one in ten are choosing not to give one or more vaccines. Some aren't giving any vaccines at all; since 1991 the percentage of unvaccinated children has more than doubled.

On the other side are doctors. Weary of parents who insist on individualized schedules, scared to send children out of their offices unvaccinated, and concerned that their waiting rooms, packed with unvaccinated children, are becoming a dangerous place, they're taking a stand. As many as four in ten pediatricians now refuse to see families who don't vaccinate, causing some parents to seek the comfort of doctors or chiropractors more willing to do what they ask.

Caught in the middle are children. Left vulnerable, they're suffering the diseases of their grandparents. Recent outbreaks of measles, mumps, whooping cough, and bacterial meningitis have caused hundreds to suffer and some to die—die because their parents feared vaccines more than the diseases they prevent.

. Amid the confusion, another group has emerged: parents angry that unvaccinated children have put their children at risk. Some of these parents have children who can't be vaccinated. Weakened by chemotherapy for their cancers, or immunosuppressive therapy for their transplants, or steroid therapy for their asthma, these children are particularly vulnerable. They depend on those around them to be vaccinated; if not, they're the ones most likely to suffer during outbreaks.

We've come to a crossroads. During the past two decades, as more states have allowed exemptions to vaccination, population immunity has broken down. And the questions have gotten harder. Should states continue to allow parents to opt out of vaccines? Or should they step in and take away that right?

The fear of vaccines, the choice to act on that fear, the consequences of that choice, and the voices rising in protest are the subjects of this book.

INTRODUCTION

All horror movies start the same way. Whether the scene is an abandoned cabin, a dark alley, or a peaceful cottage, one line of dialogue, quietly uttered five minutes before the carnage starts, is inevitable: "Did you hear something?"

Such is the case with certain infections in American children. On February 17, 2009, Robert Bazell, a science correspondent for *NBC Nightly News*, told the story of an unusual outbreak in Minnesota: a handful of children had contracted meningitis caused by the bacterium *Haemophilus influenzae* type b, or Hib. What made this outbreak so unusual was that it didn't have to happen; a vaccine to prevent Hib had been around for twenty years. But most of the Minnesota children—including one who died from the disease—weren't vaccinated. The problem wasn't that their parents couldn't afford vaccines, or that they didn't have access to medical care, or that they didn't know about the value of vaccines. The problem was that they were afraid: afraid that vaccines contained dangerous additives; or that children received too many vaccines too soon; or that vaccines caused autism, diabetes, multiple sclerosis, attention deficit disorder, learning disabilities, and hyperactivity. And despite scientific studies that should have been reassuring, many parents weren't reassured. When the outbreak was over, one mother reconsidered her decision: "The doctor looked at me and said, 'Your son is going to die. He doesn't have much time.' Honestly, I never really understood how severe the risk [was] that we put our son at."

The Minnesota outbreak wasn't an isolated event. In 2008 and 2009, outbreaks of Hib meningitis occurred in Pennsylvania, New

York, Oklahoma, and Maine, killing at least four more children; in each case, parents had made the choice not to vaccinate—a choice that proved fatal.

Most parents today have probably never heard of Hib. But older doctors certainly remember it; so do grandparents. Before the vaccine, Hib caused meningitis, bloodstream infections, and pneumonia in twenty thousand children every year, killing a thousand and leaving many with permanent brain damage. Today's outbreaks are a fraction of what they were in the past. But, as more parents choose not to vaccinate, more outbreaks of preventable infections are popping up across the country. And more children are needlessly harmed. The phenomenon doesn't seem to be going away. So the questions remain: Have we heard something? Are these outbreaks just a blip on the radar screen? Or do they represent a deeper, far more serious problem?

Hib isn't the only concern. Whooping cough, measles, and mumps—diseases once easily controlled by vaccines—are also coming back.

Whooping cough (pertussis) is a devastating infection. Before a vaccine was first used in the United States in the 1940s, about three hundred thousand cases of whooping cough caused seven thousand deaths every year, almost all in young children. Now, because of the pertussis vaccine, fewer than thirty children die every year from the disease. But times are changing.

Vashon Island is a small commuter island in King County, Washington. About ten thousand people live there, mostly well educated and affluent. The island supports an elementary school, a middle school, and a high school. Although schools on Vashon Island look like schools in any other upscale community, one difference isn't immediately obvious. About one in seven children on the island is unvaccinated; in the middle school, it's one in four. The consequences of this choice first appeared in the early 1990s. In 1994, the children of Vashon Island suffered 48 cases of whooping cough; in 1995, the number increased to 263; by 1999, it was 458.

Although whooping cough often starts benignly, it's not a benign disease. Children with whooping cough first suffer congestion, coughing, and runny noses. Then they get the symptom that gives the disease its name. The bacterium that causes whooping cough, *Bordetella pertussis*, triggers the accumulation of thick, sticky mucus in the windpipe. Children try to rid themselves of the mucus by coughing; but it's so gummy and tenacious that it's impossible to cough up. Panicking, a child coughs and coughs—as many as twenty times in a row—without breathing in. Because these coughing spells deprive children of oxygen, many cough until they're literally blue in the face. The long-awaited breath, taken against a narrowed windpipe, creates an unmistakable high-pitched sound. Parents who hear the whoop of whooping cough never forget it.

Coughing isn't the only problem. Some children with whooping cough suffer pneumonia when pertussis bacteria travel to their lungs, or seizures when their brains don't get enough oxygen, or suffocation when mucus completely blocks their windpipes. Some cough so hard that they break ribs or so long that they become malnourished.

Although Vashon Island is an excellent example of what can happen when parents stop giving pertussis vaccine, it isn't the only example. On May 10, 2008, an outbreak of whooping cough occurred at the East Bay Waldorf School in El Sobrante, California. The Waldorf School follows the teachings of Rudolf Steiner, author of *Fundamentals of Anthroposophical Medicine*. Steiner believes that vaccination "interferes with karmic development and the cycles of reincarnation." As a result of this philosophy, at least sixteen students, mostly kindergarteners, suffered the disease; virtually all were unvaccinated. When health officials investigated the outbreak—and discovered just how many children were left vulnerable—they did something that rarely occurs in twenty-first-century America. They closed the school until the epidemic subsided.

Pertussis outbreaks haven't been limited to Washington and California; they've also occurred in Delaware, Illinois, Mississippi, Arizona, Oregon, and Vermont. The outbreak in Delaware in 2006 prompted the Centers for Disease Control and Prevention (CDC) to issue a simple, frightening statement in its publication *Morbidity and Mortality Weekly Report*: "This age distribution is similar to that observed in the pre-vaccine era." One city, Ashland, Oregon, has an elementary school in which not a single child is vaccinated.

Measles is also coming back.

On May 4, 2005, a seventeen-year-old unvaccinated Indiana girl boarded a plane to Bucharest, Romania. Sent on a mission by her church, she visited an orphanage and a hospital. She didn't know that Romania was in the midst of a measles epidemic. On May 14, on the plane back to Indiana, she developed fever, cough, runny nose, and pink eye. The next day, she went to a church picnic attended by five hundred people. Although she felt ill, she was excited to share her experiences with her friends and neighbors. Neither she nor anyone at the picnic knew she had measles. On May 16, a red, speckled rash appeared on most of her body.

On May 29, two weeks after the picnic, the Indiana State Health Department received a phone call from a doctor in Cincinnati who recently had admitted a severely dehydrated six-year-old boy to the hospital. His diagnosis: measles. The doctor called health officials to tell them where the boy had been two weeks earlier: a church picnic in Indiana. The subsequent investigation was a case study in just how contagious measles virus can be. Among the 500 people at the picnic, 35 had never received a measles vaccine—31 of them (89 percent) became infected. Of the remaining 465 people, only 3 (0.6 percent) were infected. The girl who had contracted measles in Romania—after spending only a few hours in a crowd of 500 people—had managed to infect almost every person susceptible to the disease.

The Indiana outbreak was a frightening reminder of our past. Before 1963, when a vaccine was first available, measles was a common cause of suffering and death. Although most parents know that measles virus causes a rash, few know that it can also travel to the lungs and cause pneumonia or to the brain and cause inflammation (a condition called encephalitis), often resulting in seizures and brain damage. Worst of all, measles virus causes a rare disease called SSPE (subacute sclerosing panencephalitis), whereby children become progressively less able to walk, talk, or stand. Invariably they develop seizures, lapse into a coma, and die; despite heroic supportive measures, no child has ever survived SSPE. Before the vaccine, measles infected as many as four million American children, causing a hundred thousand to be hospitalized and five hundred to die every year.

Following the measles outbreak in Indiana, health officials at the CDC did everything they could to warn parents about the seriousness of the disease. They issued media alerts, health advisories, talking points, and educational materials, all with the hope that the alarm sounded in Indiana would be heard. But their warnings were ignored.

On January 13, 2008, a seven-year-old unvaccinated boy flew back to his home in San Diego after a family vacation in Switzerland. Nine days later he developed cough and a runny nose. His parents, who thought he had only a cold, sent him to school. But his illness worsened. The next day, the mother brought the boy to the doctor's office, where he sat in the waiting room with other children. The doctor, unsure of the diagnosis, sent the child to a testing laboratory at a local hospital. Later that day, the boy was taken to the hospital's emergency room with a fever of 104 degrees and a worsening rash. Because none of the attending doctors had considered measles, isolation precautions were never used in the doctor's office, the testing laboratory, or the hospital.

Between January 31 and February 19 of that year, other children started getting sick: the boy's two siblings, several of his classmates, and three children who had been sitting in the doctor's waiting room. Again, the measles virus showed its remarkable ability to find susceptible children. Every child infected by the boy was unvaccinated. Of the three children who caught measles in the doctor's waiting room, all were too young to have been vaccinated; one was hospitalized with severe dehydration; another traveled by plane to Hawaii while contagious. The measles outbreak in California shouldn't have been surprising. In 2008, the parents of ten thousand California kindergarteners chose not to vaccinate their children.

California wasn't the only state to suffer a measles outbreak. Thirteen other states—Illinois, Washington, Arizona, Hawaii, Wisconsin, Michigan, Arkansas, Georgia, Louisiana, Missouri, New Mexico, Pennsylvania, and Virginia—as well as the District of Columbia succumbed. When it was over, 140 children, almost all of them unvaccinated, had been infected; 20 were hospitalized. It was the largest single measles outbreak in the United States in more than a decade.

The Indiana and nationwide outbreaks shared one important feature: in both cases, the first infection occurred outside the United States. This isn't unusual. Every year about sixty people traveling from countries where immunization rates are lower, such as Switzerland, Austria, Ireland, Israel, the Netherlands, Japan, and the United Kingdom, enter the United States with measles. Indeed, all of these countries continue to suffer measles outbreaks. But the situation in 2008 was different; this time measles spread from one unvaccinated American child to another to another. The problem wasn't that national immunization rates were low; they were actually quite high. The problem was that certain communities had so many unvaccinated children that infections could spread unchecked.

Perhaps most disturbing was an outbreak of mumps among Hasidic Jews in New York and New Jersey—an outbreak that showed just how much we depend on one another for protection.

In June 2009, an eleven-year-old boy traveled to England and caught the mumps. At the time, thousands of British children were infected with mumps, primarily because their parents were afraid that the measles-mumps-rubella (MMR) vaccine caused autism. On June 17, the boy flew back to New York, attended a summer camp for Hasidic Jews, and started a massive epidemic. By October, two hundred people had been infected; by November, five hundred; and by January 2010, fifteen hundred. When it was over, mumps was found to have caused pancreatitis, meningitis, deafness, facial paralysis, or inflammation of the ovaries in sixty-five people; nineteen were hospitalized.

The mumps outbreak in 2009 showed that even vaccinated people are at risk. In order to stop the spread of infections, a certain percentage of the population needs to be vaccinated, a phenomenon known as population or herd immunity. People who aren't vaccinated or who can't be vaccinated will be protected when surrounded by a highly vaccinated group, much like a moat safeguards a castle. The fraction of the population that needs to be vaccinated to provide herd immunity depends on the contagiousness of the infection. For highly contagious infections—such as measles or pertussis—the immunization rate needs to be about 95 percent. For somewhat less contagious infections—like mumps and rubella—herd immunity can be achieved with immunization rates around 85 percent. Although 70 percent of the Hasidic Jews in the mumps outbreak of 2009 were vaccinated, the proportion of those protected was actually lower. That's because no vaccine is 100 percent effective. For mumps, about 88 percent are protected after two doses. Therefore, although 70 percent were immunized, only 62 percent were protected, well below the rate needed to stop the spread of mumps.

The epidemic among Hasidic Jews wasn't an isolated event. Three years earlier, in 2006, mumps had swept across the Midwest, infecting more than sixty-five hundred people, mostly college students.

The mumps epidemics of 2006 and 2009 proved that even vaccinated people might not be protected if vaccination rates aren't high enough.

Outbreaks started by travel outside the United States won't be limited to measles and mumps.

In 2003, rumors circulated in Nigeria that polio vaccine caused AIDS and made young girls infertile. Vaccination programs came to a halt. By 2006, polio originating in Nigeria had spread to twenty previously polio-free countries in Africa and Asia—more than five thousand people were severely and permanently paralyzed. "Such large-scale polio outbreaks haven't been seen in quite a long time," said Tammam Aloudat, a senior health official at the International Red Cross. Walter Orenstein, deputy director for the Gates Foundations' Global Health Program, saw a parallel with the U.S. measles outbreaks of 2008. "Polio is only a plane ride away from the United States," said Orenstein. "If we let our guard down, if our immunization coverage drops, there is certainly the possibility of a polio outbreak." Orenstein doesn't think it will end with polio: "Diphtheria could also come back. Any of them can. Because aside from smallpox, every other one of these infections is either in the United States or close to our borders."

In the early 1900s, children routinely suffered and died from diseases now easily prevented by vaccines. Americans could expect that every year diphtheria would kill twelve thousand people, mostly young children; rubella (German measles) would cause as many as twenty thousand babies to be born blind, deaf, or mentally disabled; polio would permanently paralyze fifteen thousand children and kill a thousand; and mumps would be a common cause of deafness. Because of vaccines, all these diseases have been completely or virtually eliminated. But now, because more and more parents are choosing not to vaccinate their children, some of these diseases are coming back.

How did we get here? How did we come to believe that vaccines, rather than saving our lives, are something to fear? The answer to that question is rooted in one of the most powerful citizen activist groups in American history; founded in 1982, it is a group that, despite recent epidemics and deaths, has continued to gain followers in both the United States and the world.

CHAPTER 1

The Birth of Fear

If you say in the first chapter that there is a rifle hanging on the wall, in the second or third chapter it absolutely must go off.

—S. SHCHUKIN, *MEMOIRS* (1911)

Frederick Wiseman was born on January 1, 1930. After graduating from Williams College and Yale University Law School, Wiseman became a law professor at Boston University. Then he decided to make movies. For the next thirty years Frederick Wiseman was the most inventive, most reviled, most controversial, and most influential documentary filmmaker in America.

Wiseman's first film—released in 1967—was his most powerful. Called *Titicut Follies*, it was a stark depiction of life inside the walls of Bridgewater State Hospital for the Criminally Insane. Wiseman showed prisoners being hosed down, force-fed, and tortured by an indifferent, bullying staff. In one scene, a physician takes a long tube and inserts it into a prisoner's nose. Then he attaches the tube to a funnel, fills it with thick, dark fluid, and stands precariously on a chair, a cigarette dangling from his mouth. A guard mockingly shouts, "Chew your food, Joey." The viewer is at once sickened by the degradation of force-feeding and captivated by the cigarette ash dangling over the funnel.

Time magazine called *Titicut Follies* a "relentless exposé of a present-day snake pit." Vincent Canby of the *New York Times*

1

wrote that the film made "*Marat/Sade* look like *Holiday on Ice.*"
And one theatrical poster warned, "Don't turn your back on this
film . . . if you value your mind or your life." *Titicut Follies* was so
hard to watch—so unblinking, so unsettling, so unfailingly detailed—
that days before its debut at the New York Film Festival, Massachusetts Superior Court judge Harry Kalus ordered the state to seize
all copies, writing: "No amount of rhetoric, no shibboleths of 'free
speech' and the 'right of the public to know' can obscure or mas-
querade this pictorial performance for what it really is—a piece of
abject commercialism, trafficking in the loneliness, on the human
misery, degradation and sordidness in the lives of these unfortunate
humans." In 1968, *Titicut Follies* was the first and only film in the
United States ever to be banned for reasons other than obscenity or
national security. Twenty years would pass before the movie was
shown to the American public.

The modern American anti-vaccine movement was born on April
19, 1982, when WRC-TV, a local NBC affiliate in Washington,
D.C., aired a one-hour documentary titled *DPT: Vaccine Roulette*.
Although Frederick Wiseman wasn't involved in a single aspect of
the film, his influence on the writer and producer, Lea Thompson,
was apparent. *Vaccine Roulette* contained the sad, haunting images
of *Titicut Follies*, except that this time, instead of inmates degraded
by prison guards, the camera focused on children—twisted, withered, disabled children—irreparably damaged by a vaccine. (Although Lea Thompson referred to the vaccine as DPT, doctors called
it DTP because the vial read "diphtheria and tetanus toxoids and
pertussis vaccine.")

   Vaccine Roulette opens with Lea Thompson standing in the middle of a newsroom, staring straight into the camera. Her tone is
grim, her voice unwavering. "DPT," she begins, "the initials stand
for diphtheria, pertussis, tetanus: three diseases against which every
child is vaccinated. For more than a year we have been investigating 'P': the pertussis portion of the vaccine. What we have found
are serious questions about the safety and effectiveness of the shot.
The overriding policy of the medical establishment has been to ag-

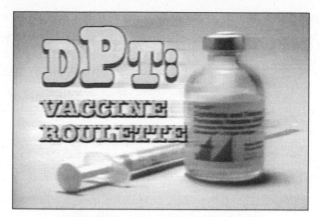

DPT: Vaccine Roulette, which aired on April 19, 1982, ignited America's modern anti-vaccine movement. (Courtesy of WRC-TV/NBC News.)

gressively promote the use of the vaccine, but it has been anything but aggressive in dealing with the consequences. Our job in the next hour is to provide enough information so that there can be an informed discussion about this important subject. It affects every single family in America."

The next image is that of a baby screaming: a needle jabbed into her arm. "It's a fact of life," says Thompson. "All children must get four DPT shots to go to school. Shots we are told will keep our children healthy. Shots we are told will protect every child from a dreaded disease: pertussis. But the DPT shot can also damage to a devastating degree." To the sound of a beating heart, the screen fills with images of children with severe mental and physical handicaps, withered arms and legs, gazing at the ceiling, drooling, seizing. Then a vial of vaccine appears behind letters spelling out "D-P-T: V-a-c-c-i-n-e R-o-u-l-e-t-t-e," each letter accompanied by a sharp, penetrating noise, like gunshots.

"The controversy isn't really over the fact that [brain damage] happens," says Thompson, "but how often it happens and whether it happens often enough to deem the vaccine more dangerous than the disease itself. You don't have to ask the Grants of Beaver Dam,

Wisconsin, that question." The next scene features a young man with emaciated, spastic legs, rhythmically shaking his head back and forth. A graphic describes his problem: "SCOTT GRANT, AGE 21, REACTION: HARSH CRY, INFANTILE SPASMS, SEVERELY DISABLED, RETARDED." Scott's mother, Marge, explains what happened to her son: "We had a child up to four months of age that was developing beautifully well. The doctor explained that he was giving Scott his first DPT shot. Between 12 and 14 hours [later], he gave an outburst of a very hard cry. What we learned later were infantile spasms [a form of epilepsy]. I went home and cried. Jim cried. We couldn't believe that we could possibly have such a black future."

"I had to start a business for myself," said Jim Grant. "I had to be home all the time in regards to helping lift him and take care of his many needs. It's quite a big job. We have not had a vacation for twenty-one years. We simply can't go away. It's impossible to go away."

Other brain-damaged children appear—all staring blankly, all clearly struggling, all allegedly harmed by the pertussis vaccine:

"POLLY GAUGERT, AGE 7, REACTION: FEVER, UNCONTROLLED SEIZURES, BRAIN DAMAGE." "I said that maybe she should not have this shot because it seems to me she was not quite herself," recalled Polly's mother. "And [the doctor] checked her all over and he said, 'She looks okay to me,' and then he gave her the shot. And the next morning when I was feeding her she went into a grand mal seizure. . . . I didn't know what was happening. I thought she was dying in my arms."

"ABRA YANKOVICH, AGE 2, REACTION: STOPPED BREATHING, SEIZURES, SEVERELY DISABLED, RETARDED." "When she was four months old, on the same day that she had her vaccination, she had her first seizure," said Abra's mother. "She was shaking and she was turning blue and she appeared to have breathing problems. By the time we got her to the emergency room she was okay. And we told her doctor that she had had her vaccination that day. Could there be a link? He said no, she was probably just choking. Just take her home and she'll be fine. But two weeks later she went into a grand mal seizure. She was very near dying." The Yankoviches visited a pediatric neurologist in Chicago where they learned Abra's fate: "We've been told that she probably will never walk on her own and she probably will never talk."

"ANTHONY RESCINITI, AGE 19, REACTION: PERSISTENT CRY, FEVER, SEIZURES, SEVERELY DISABLED, RETARDED." "Tony Resciniti, 19 years old: he suffers a convulsion about once a day," says Thompson. "The drugs to control [the convulsions] cost $1,200 a year. Tony convulsed within twenty-four hours of getting the DPT shot." Tony wasn't the only one in his family to suffer from DTP.

"LEO RESCINITI, AGE 17, REACTION: FEVER, CONVUL-SIONS, SEVERELY DISABLED, RETARDED." Thompson: "Leo Resciniti, seventeen years old: only a few hours after his first DPT shot, Leo, too, went into convulsions. His temperature soared."

"KELLI HOLCOMB, AGE 8, REACTION: PERSISTENT CRY, STIFFNESS, SPASTIC QUADRIPLEGIC, BRAIN DAMAGE." "Kelli Holcomb got her shots from the U.S. Army," says Thompson. "Her parents were told nothing of the risks of the DPT vaccine."

Doctors also weigh in. Robert Mendelsohn, a pediatrician from Chicago, says, "It's probably the poorest and most dangerous vaccine that we now have, [and] the dangers are far greater than any doctors have been willing to admit." Gordon Stewart, an epidemiologist from Scotland, says, "I believe that the risk of damage from the vaccine is now greater than the risk of damage from the disease." Jerome Murphy, a pediatric neurologist from Milwaukee, says, "There is overwhelming data that there is an association. I know it has influenced many pediatric neurologists not to have their children immunized with pertussis." One father recalls, "Dr. Millichap told us . . . personally, he wouldn't even give that [vaccine] to his dog."

Then Thompson reveals something even more disturbing: doctors had known about the horrors of pertussis vaccine for decades. "Medical knowledge about severe reactions to the whooping cough vaccine goes all the way back to the early '30s," she says. "The *Pediatric Red Book*, written by the American Academy of Pediatrics, lists high fever, collapse, shock-like collapse, inconsolable crying, convulsions, and brain damage as reactions to the DPT vaccine. Those complications are associated with varying degrees of retardation, ranging from severe brain damage, like Scott, to learning disabilities."

Thompson ends her show on an ominous note. A young boy, having just received a vaccine, is screaming. Terrified, he reaches for his father who tries to reassure him: "It's all right. It's one of those things that little boys have to have. See, it's all gone already." The sound of a heartbeat grows louder. The little boy looks straight into the camera, worry creasing his face. It's as if he knows that the shot isn't "all gone"—that future horrors await him.

Vaccine Roulette aired two more times in Washington, D.C., and nationally on *The Today Show*; within weeks, magazines and newspapers across the country told the stories of children permanently damaged by the pertussis vaccine.

Physicians were stricken. Leonard Rome, a pediatrician in Shaker Heights, Ohio, said the program was "devastating in every pediatrician's office. Doctors were calling each other and saying, 'Are you still giving the pertussis [vaccine]?'" In New Mexico, Dr. James Waltner said, "Inquiries about the vaccine have increased 25 percent." And on the West Coast, Robert Meechan, a professor of pediatrics at the Oregon Health Sciences University in Portland, said, "We have had a lot more explaining to do." Thousands of parents called their doctors to reject the pertussis vaccine or to report a laundry list of side effects. Many questioned all vaccines; and some, the integrity of those who gave them.

Vaccine Roulette started a firestorm. "The WRC-TV switchboard was melting from all of the calls coming in," recalled the CDC's Alan Hinman. During the furor, several parents got together and decided to do something about it—to take control of a situation that appeared out of control. The organization they formed would forever change how American parents thought about vaccines.

Kathi Williams was twenty-seven years old when she watched *Vaccine Roulette* in her one-bedroom apartment in Fairfax, Virginia. "I had taken my son into the doctor's office for his fourth DPT shot," she recalled. "I was a very well-educated parent. I'd read every book on childcare, childbirth, nursing; whatever book was out there I read. [But] there was never one word about vaccinations and vaccine problems. So I was horrified when I saw this show because four days prior to that my very happy, healthy, beautiful bouncy boy, who never cried, screamed his head off for over eight hours. It was a high-pitched, uncontrollable scream. In between the periods of the high-pitched crying he would fall into a very deep sleep. And then he would just wake up again like someone had pinched him and start this screaming again. My doctor told me that it was normal." Williams asked her mother what she should do. "She said, 'Call Lea Thompson.'"

Jeff Schwartz and his wife Donna Middlehurst watched the show from their home in Silver Spring, Maryland. Schwartz was an environmental lawyer and Middlehurst a securities lawyer. Their daughter, Julie, had received her third DTP shot in July 1981, nine months before *Vaccine Roulette* aired. On the afternoon of her shot, Jeff was holding his daughter when he noticed "a sort of startle." The startle turned into a seizure that lasted forty minutes. Other seizures followed; when they were finally under control, Schwartz posed a question: "We asked the doctor about the DTP. And she said, 'No, it's actually fevers that produce these things.'" But after watching *Vaccine Roulette*, Schwartz and Middlehurst knew differently. "We said, 'Oh my God. Now we know what happened.'" On March 25, 1984, Julie Schwartz died during a seizure. Later, Jeff lamented the irony of the DTP vaccine: "To take your daughter in to protect her and have that be the agent that destroys her."

Barbara Loe Fisher was thirty-four years old when she watched a rebroadcast of *Vaccine Roulette* the day after it first aired. Fisher was the mother of a four-year-old son, Christian. She remembered what happened the night after he received his fourth DTP shot: "Several hours after we got home, I realized how quiet it was in the house and went upstairs to look for Chris. I walked into his bedroom to find him sitting in a rocking chair staring straight ahead, as if he couldn't see me standing in the doorway. His face was white and his lips were slightly blue. When I called out his name, his eye-lids fluttered; his eyes rolled back in his head; and his head fell to his shoulder. It was as if he had suddenly fallen asleep sitting up. When I picked him up and carried him to his bed, he was like a dead weight in my arms. In the following days and weeks, Chris deteriorated. He no longer knew his alphabet or numbers, and he couldn't identify the cards he once knew so well. He couldn't concentrate for more than a few seconds at a time. My little boy, once so happy-go-lucky, no longer smiled." Fisher called WRC-TV and was given the name of Kathi Williams. Then she drove to Williams's house and, in 1982, started an organization that has lasted for three decades. They called themselves Dissatisfied Parents Together (DPT).

A few years after Kathi Williams, Jeff Schwartz, and Barbara Loe Fisher formed Dissatisfied Parents Together, Fisher became its president. Well suited to the role of public spokesperson, Fisher had received a degree in English from the University of Maryland, served as an editor for the New York Life Insurance Company in New York City, and coordinated media relations for the Alexandria, Virginia, tourist council.

Fisher's anger was white-hot and rarely contained. She simply couldn't stand that children were forced to get vaccines—forced by a government that licensed, recommended, and mandated them; forced by public health officials who were asleep at the switch and, worse, didn't seem to care; and forced by pharmaceutical companies with little interest in making vaccines safer. Barbara Loe Fisher's anger at a medical establishment that had required her to vaccinate her son never subsided. By the early 1990s, Dissatisfied Parents Together had changed its name to the National Vaccine Information Center, the single most powerful anti-vaccine organization in America. For the next three decades, Fisher would use that anger to try to convince a generation of American parents that vaccines were far more dangerous than they'd realized.

On May 7, 1982, Senator Paula Hawkins, a Republican from Florida, called a hearing before the Committee on Labor and Human

Paula Hawkins, a Republican senator from Florida, chaired a congressional hearing to evaluate possible brain damage caused by pertussis vaccine. (Courtesy of Paul Hosefros/ *New York Times*.)

Resources of the U.S. Senate. Only eighteen days had elapsed since the airing of *Vaccine Roulette*. The speed of the Hawkins hearing was the result of a series of chance events. Lea Thompson had first become interested in pertussis vaccine after she'd been contacted by the parents of Tony and Leo Resciniti, the teenagers from New York who had apparently suffered permanent brain damage. The Rescinitis, as it turned out, were cousins of Dan Mica, a Republican congressman from Florida. On April 28, 1982, nine days before the Hawkins hearing, Kathi Williams, Jeff Schwartz, Barbara Loe Fisher, and several other parents met to discuss strategy at Dan Mica's office in Washington, D.C. Mica's brother, John, was on Paula Hawkins's staff.

Hawkins opened her hearing with a statement. "The immunization program is now threatened on another front," she warned: "the fear of adverse health events resulting from immunization. To combat this fear and to achieve and maintain high immunization rates, full public communication and health education is essential. The general public has a right to be given information about vaccines— even in areas of scientific or medical uncertainty." Hawkins then made an ominous and all-too-accurate prediction of future events. "It would be tragic if efforts to eliminate or control communicable disease were to become hampered because the public's confidence was so eroded as to cause frightened segments of the population to oppose and reject vaccines. Neither can we afford revival of serious childhood epidemics because a complacent and apathetic public, with a diminishing memory, forgets the iron lung."

One of the first parents to testify was Kathi Williams. On behalf of Dissatisfied Parents Together, she made a list of demands. "Number one: Although several studies have been done, why has the government had a limited research program dealing with adverse effects of vaccines? Number two: Why hasn't a safer vaccine been developed? Number three: Why haven't high-risk children been identified? Number four: Why haven't physicians been required to report adverse reactions to a central recordkeeping agency? Number five: Why haven't physicians and parents been better informed about the possible reactions to the pertussis vaccine? Number six: Should the states mandate that the present pertussis vaccine be given to all children who attend school? Number seven: Should there be a compensation program for children who have been retarded or seriously disabled by the pertussis vaccine?"

Remarkably, within a few years, almost all of Kathi Williams's demands would be met.

Pediatricians used the Hawkins hearing as a chance to attack *Vaccine Roulette*. In a written statement, the American Academy of Pediatrics (AAP) called Thompson's program "unbalanced," "biased," "inaccurate," and "superficial," and claimed it "unnecessarily frightened laypersons." CDC officials complained that *Vaccine Roulette* had dismissed the seriousness of whooping cough, unfairly characterized doctors and health officials as ignorant of the vaccine's side effects, and inaccurately claimed that the vaccine didn't work very well. But despite their criticisms, not a single physician who testified at Hawkins's hearing disagreed with Lea Thompson's most damning accusation—that the "P" in the DTP vaccine had caused permanent harm. Edward Mortimer, professor of pediatrics at Case Western Reserve University and probably the most recognized vaccine expert in the United States, said, "Our best estimates are that, of the three and a half million children born annually in the United States, between twenty and thirty-five incur permanent brain damage as a result of the vaccine. Each of us concerned with vaccine recommendations believes that this is twenty to thirty-five kids too many."

For years doctors had argued that the benefits of the pertussis vaccine outweighed its risks. Now, because of one television program, the public's perception of those risks was tipping in the other direction. Thousands of parents were choosing not to vaccinate their children.

It was only the beginning.

Lea Thompson's career was meteoric. After *Vaccine Roulette*, she worked as a contributing correspondent to NBC's *Today Show* and *NBC Nightly News with Tom Brokaw*, produced and hosted a weekly half-hour magazine called *Byline: Lea Thompson*, and worked as the chief consumer correspondent at *Dateline NBC* and *MSNBC*. For her investigative reporting she won almost every major award in broadcasting, including two Peabody awards, two Polk awards, a Columbia Dupont award, a Loeb award, a National Emmy, the Edward R. Murrow award, multiple National Headliner awards, National Press Club awards, and two dozen Washington Regional Emmys. She was named Washingtonian of the Year in 1989. And her reporting made a difference. As a result of her stories, unsafe toys have been removed from shelves, millions of hairdryers containing asbestos have been recalled, procedures at Sears now ensure that old batteries aren't sold as new, grocery stores have adopted policies for checking ground beef, warning labels have been placed on Infant Tylenol, and the largest manufacturer of defibrillators in the United States has been shut down. But no story had a greater impact than *Vaccine Roulette*. If the government hadn't stepped in several years later, Thompson's show could have eliminated vaccines from the American marketplace.

Fifteen years later, while receiving an award from the National Vaccine Information Center, the group she had essentially founded,

Thompson said, "*DPT: Vaccine Roulette* stands out as one of the most important stories of my life—maybe *the* most important story of my life. And I can tell you that I only have one regret—only one. And that's that we didn't do this story ten years earlier. That we didn't know about it ten years earlier. Because so many kids might not have suffered and so many kids might still be alive."

Thompson's program created one of the most powerful advocacy groups in American history. But Lea Thompson wasn't the first investigative journalist to report problems with the pertussis vaccine; and the United States wasn't the birthplace of the modern-day antivaccine movement. All of these events had already occurred—eight years earlier in England. Indeed, it was the concern of British parents that led to a study that prompted physicians like Jerome Murphy, Gordon Stewart, and Robert Mendelsohn to claim that the pertussis vaccine caused permanent brain damage—a claim that two decades later was found to be wholly and utterly incorrect.

CHAPTER 2

This England

This blessed plot, this earth, this realm, this England.
—WILLIAM SHAKESPEARE,
KING RICHARD II

On Friday, October 26, 1973, John Wilson, a pediatric neurologist, stood in front of a group of professors, consultants, and specialists at the Royal Society of Medicine in London. Wilson placed a typed manuscript on the lectern and looked up. What he was about to say arguably would lead to more suffering, more hospitalizations, more permanent disabilities, and more deaths than any other pronouncement in the history of vaccines.

"Between January 1961 and December 1972," Wilson began, "approximately fifty children have been seen at the Hospital for Sick Children in London because of neurological illness thought to be due to the DTP inoculation." For years Wilson had accumulated these children's stories. For years he had struggled with the damage caused by pertussis vaccine. Now it was time to tell the world about it. Wilson reported one child who had transient blindness and mental deterioration. Another had vomited for four days, become blind, and died six months later during an uncontrolled seizure. Yet another had been completely paralyzed on one side of her body. The final accounting was grim: of fifty children studied, twenty-two had become mentally disabled or epileptic or both. To John Wilson, the cause of all of this suffering and death was clear. "We do not think . . . that the majority of cases here represent a chance association," he said. Wilson was convinced by "the clustering of illness in the seven days after inoculation and particularly in the first 24 hours" that the damage had been caused by pertussis vaccine.

Although he was the most influential, John Wilson wasn't the first to propose that pertussis vaccine permanently harmed or killed children.

In 1933 Thorvald Madsen from the State Serum Institute in Denmark reported two children who had died after receiving pertussis vaccine.

In 1946 Jacob Werne and Irene Garrow from St. John's Long Island City Hospital in New York reported twin brothers who "cried considerably," "vomited," "fell asleep," and "when next noticed by their parents, appeared 'lifeless.'" One child was dead on arrival to the hospital; the other died a few hours later.

In 1948 Randolph Byers and Frederic Moll, from Boston Children's Hospital and Harvard Medical School, reported fifteen children with seizures, coma, or paralysis within a day of receiving pertussis vaccine. Most became severely retarded; two died.

In 1960 Justus Ström from the Hospital for Infectious Diseases in Stockholm, Sweden, reported thirty-six children harmed by pertussis vaccine. "In twenty-four of these, the initial symptom was convulsions, in six cases coma, and in four, acute collapse." Seven years later, Ström examined the records of more than five hundred thousand children, this time finding one hundred and seventy with seizures or "destructive brain dysfunction" or shock.

In 1973, when John Wilson finished his presentation to the Royal Society of Medicine a murmur spread through the crowd. The society was one of the most prestigious institutions in London, and Wilson worked at the Hospital for Sick Children at Great Ormond Street, a world-renowned medical center. Further, Wilson, a doctor of philosophy and medicine, was a member of the prestigious Royal College of Physicians. When John Wilson said that pertussis vaccine caused brain damage, the charge was taken seriously.

Six months after his presentation to the Royal Society—and three months after his study was published—John Wilson appeared on a thirty-minute prime-time British television program called *This Week*. A precursor to Lea Thompson's *DPT: Vaccine Roulette*, the program featured frightening images of children allegedly damaged by pertussis vaccine—and terrifying calculations of how any child could be one of them. One caption read, "Every year about 100 are brain damaged." Six minutes into the program, John Wilson appeared. Now, instead of influencing a handful of specialists at the Royal Society, or a few thousand readers of a medical journal, he was talking to millions of television viewers. Wilson was asked whether he was convinced that pertussis vaccine caused permanent harm. "I personally am," he replied. "Because now I've seen too many children in whom there has been a very close association between a severe illness, with fits, unconsciousness, often focal neurological signs, and inoculation." The reporter asked him what he meant by "too many children." Wilson, recalling his experiences at Great Ormond Street Hospital, replied, "Well, in my time here, the last eight and a half years, I personally have seen somewhere in the region of eighty patients."

The media exploded. Headlines in influential British newspapers read, "Whooping Cough Vaccine Risks Concealed, Say Victims' Parents," "Vaccine Call Is Attacked," "Whooping Cough Vaccine 'Should Be Abandoned,'" "New Campaign to Win State Help for the Vaccine-Damaged," "Dangers of a Shot in the Dark," and "Boy's Brain Damaged in Vaccine Experiment."

Doctors also sounded the alarm. George Dick, a respected microbiologist at Queens University in Belfast, said, "I would not recommend the vaccine for infants living in communities where there is good maternal and medical care." Gordon Stewart, the epidemiologist from the University of Glasgow who later appeared on *Vaccine Roulette*, said, "The Department of Health and Social Security . . . refuse[s] to acknowledge brain damage as anything but a doubtful and rare consequence of vaccination. The facts suggest otherwise." David Kerridge, a professor of statistics at Aberdeen University, joined Stewart in decrying the vaccine. "My advice would be to abandon vaccination," he said.

Health officials in England waited for the other shoe to drop. In October 1975, an editorial published in the *British Medical Journal* warned: "A sharp decrease in the number of infants routinely immunized after the 1974 adverse publicity could mean that there will be a substantial pool of susceptible children when the next epidemic is due in 1978. Will this mean that there will then be a resurgence of whooping cough?" The answer came quickly. The year before Wilson's paper, 79 percent of British children were immunized. By 1977, the rate had fallen to 31 percent. As a consequence, more than a hundred thousand children contracted whooping cough; five thousand were hospitalized; two hundred had severe pneumonia; eighty suffered seizures; and thirty-six died. It was one of the worst epidemics of whooping cough in modern history. Dr. David Salisbury, later to be director of immunization at England's Department of Immunization, was a young pediatrician when the epidemic hit: "At a practical level, I was very aware of the drop in coverage of the pertussis vaccine. Every evening that I was on duty children were referred to the hospital with whooping cough. I remember ventilating them for weeks and weeks and weeks and there were some that we simply couldn't get off the ventilators."

Fear of pertussis vaccine spread. In Japan, after health officials placed a moratorium on the vaccine, the number of hospitalizations and deaths from pertussis increased tenfold.

John Wilson later became an advisor to the Association of Parents of Vaccine Damaged Children and its founder, Rosemary Fox, a forty-six-year-old social worker. By advising Fox, Wilson had completed the cycle of events in London that would be repeated in Washington, D.C., several years later: a national television program alerting parents to the harm of pertussis vaccine; formation of a parents' group seeking compensation; a media outcry in support of aggrieved parents; and irreparable, unending suspicion that vaccines were doing far more harm than good.

Probably no one was in a better position to comment on events in England than Dr. James Cherry. Currently a professor of pediatrics at UCLA's School of Medicine, Cherry has co-authored the leading textbook on pediatric infectious diseases, published hundreds of articles and book chapters on pertussis and pertussis vaccine, lectured before physicians and scientists internationally, and won many prestigious awards. James Cherry is arguably the world's expert on pertussis.

Soon after John Wilson issued his terrifying report, Cherry moved to England to study pertussis at the London School of Hygiene and Tropical Medicine. He noticed one critical difference between the United States, where immunization rates dropped only slightly, and England, where immunization rates dropped precipitously. "It wasn't the public; it was the doctors," recalled Cherry. "It was the family physicians who really stopped vaccinating." (A survey by the London *Sunday Times* in 1977 found that 47 percent of general practitioners "would not recommend" the pertussis vaccine for their patients.) Further, Cherry found that the number of children who died from whooping cough was vastly underestimated. "I noticed the pertussis deaths in the epidemic . . . were incredibly low," he recalled. Cherry knew the reason why: physicians weren't accurately reporting the disease. "[Doctors] didn't want the parents to have guilt feelings but it turned out [the doctors] didn't want to have guilt feelings. So they were reporting [pertussis deaths] as other things." Cherry found that pertussis hadn't killed thirty-six British children (as was officially reported); it had killed six hundred. Most were classified as respiratory deaths without mention of pertussis, and two hundred pertussis deaths were classified as Sudden Infant Death Syndrome (SIDS). "SIDS cases went up during the epidemic," said Cherry.

British health officials were in a tough spot. They couldn't claim that the benefits of pertussis vaccine outweighed its risks when risk estimates were all over the map. So they decided to fund a study that would determine once and for all the risk of brain damage from pertussis vaccine. Then and only then could parents weigh the risk of getting the vaccine against the risk of not getting it.

To assess the risks, health officials turned to Dr. David Miller, a professor of community medicine at the Central Middlesex Hospital in London. Miller and his colleagues launched the most comprehensive, expensive, and time-consuming study to date. Between

1976 and 1979, Miller's team asked consultant pediatricians, infectious disease specialists, and neurosurgeons to report any children who had serious neurological illnesses, then determined whether those children were more likely to have recently received DTP than normal children. Miller found "a statistically significant association with diphtheria, tetanus, and pertussis vaccine . . . especially within 72 hours." According to the Miller study, DTP caused permanent brain damage in one in one hundred thousand children given three doses of vaccine.

Miller's study was the first to address the question of risk using appropriate controls. As a consequence, academic physicians around the world believed it. When vaccine expert Edward Mortimer stood in front of Paula Hawkins's committee in 1982 and declared pertussis vaccine to be a rare cause of permanent harm, it was David Miller's study he was thinking about.

In the United States, the dominoes fell: parents decried the vaccine, the media trumpeted their claims in dramatic headlines, and medical experts supported them with evidence from Miller's study. It was a perfect storm. And it added up to one thing: lawsuits. Many, many lawsuits.

During *Vaccine Roulette*, Lea Thompson offered a preview of coming events. "More and more families of DPT victims are deciding to sue," said Thompson. "Not only doctors, but manufacturers and the government." Personal-injury lawyers advertised their services on television and radio commercials, in newspapers, in magazines, and on the backs of telephone books. They urged parents of vaccine-damaged children to come forward, to get the justice and compensation they deserved. In 1981, one year before *Vaccine Roulette*, 3 lawsuits were filed against vaccine makers. By the end of 1982, lawyers had filed 17 lawsuits; during each of the next four years, they filed 41, 73, 219, and 255.

Jurors were sympathetic. On March 7, 1983, four-month-old Tyler White "suffered a seizure lasting several hours" after his second DTP vaccine. A few months later he had another, then another, and was diagnosed with epilepsy and severe developmental delays.

Three years later, a jury awarded Tyler $2.1 million. On March 17, 1980, Michelle Graham "developed a severe and irreversible neurological condition known as encephalopathy [brain damage]" after her first DTP shot. The jury awarded Michelle $15 million. Melanie Tom received $7.5 million. Other lawsuits included awards for $5.5 million, $2.5 million, and $1.7 million, and many more were settled out of court "in the million-dollar range."

The amount of money requested by plaintiffs increased exponentially from $25 million in 1981 (one year before *Vaccine Roulette*) to $414 million in 1982, $655 million in 1983, $1.3 billion in 1984, and $3.2 billion in 1985. In response, pharmaceutical companies increased the prices of their vaccines and scrambled to get liability insurance. In early 1982, DTP vaccine cost $0.12 per dose. In June 1983, the cost rose to $2.30; the next year, to $2.80. By 1985, the cost of one dose of DTP vaccine was $4.29—a thirty-five-fold increase in less than three years. Increased revenues didn't offset the cost of awards. In 1984, the amount claimed in lawsuits exceeded DTP sales twentyfold. In 1985, despite a near doubling of the price, the damages claimed exceeded sales thirtyfold.

The result was predictable. Pharmaceutical companies abandoned vaccines. In 1960, seven companies made DTP. By 1982 only three remained: Connaught Laboratories of Swiftwater, Pennsylvania; Lederle Laboratories of Pearl River, New York; and Wyeth Laboratories of Philadelphia. On June 13, 1984, Wyeth announced it would no longer be distributing DTP. Later that summer, Connaught announced it was unable to get liability insurance and would stop making DTP vaccine for American children. Following Connaught's announcement, Lederle was the only company left standing.

On December 19, 1984, James O. Mason, director of the Centers for Disease Control and Prevention, appeared before the House Subcommittee on Health and the Environment. The committee wanted to know how much DTP vaccine was available. The situation, as Mason described it, was desperate. "On November 27, [1984,] Lederle informed us that they were having some production difficulties and that two production lots scheduled for release in

January and February 1985 would not be available. Contacts with state health departments were immediately undertaken and it was determined that there were approximately 1.5 million doses on hand in the states." Mason then resorted to understatement: "Comparing this amount to an average national monthly use . . . indicates that vaccine supplies would be essentially exhausted before the end of February 1985." In three months, the United States would run out of pertussis vaccine. Mason knew what was at stake; he had to do something to extend the supply. So he recommended a suboptimal vaccine schedule, figuring that some immunity was better than no immunity: "[We have] developed recommendations to try to ensure maximum prevention during the period of likely shortage. These involved delaying administration of the fourth dose of DTP, usually given at eighteen months of age, and the fifth dose, usually given at four to six years of age."

On February 12, 1985, a few months after the CDC had recommended withholding the fourth and fifth doses of DTP vaccine, the American Academy of Pediatrics held an emergency meeting to discuss vaccine shortages. Representatives from the American Medical Association, the American Academy of Family Physicians, the Department of Defense, the Department of Health and Human Services, pharmaceutical companies, and state, county, and city health departments attended. The news wasn't good. A survey of hundreds of physicians found that although most had followed the CDC recommendation, one in three still couldn't find enough vaccine.

It got worse.

In 1979, three-month-old Kevin Toner became permanently paralyzed from the waist down after receiving DTP. Kevin suffered an uncommon disorder called transverse myelitis, in which one segment of the spinal cord becomes inflamed. There was then and remains now no evidence that either pertussis or pertussis vaccine causes transverse myelitis. But in a courtroom, that didn't matter. The jury awarded Kevin $1.13 million. The company that was sued was Lederle Laboratories—the only American company still distributing pertussis vaccine. To Lederle, the message was clear. It wasn't only children with epilepsy and mental retardation who could be compensated. Everything was on the table. Lederle knew that its vaccine prevented only whooping cough, tetanus, and diphtheria, not every other illness that occurred in the first year of life. The Toner case was the last straw. On April 1, 1986, Lederle Laboratories announced to the AAP and the Department of Health and Human Services that it would no longer produce and distribute DTP vaccine.

Other vaccines suffered. The number of companies making measles vaccine dropped from six to one and those making polio vaccine from three to one. Vaccine makers were getting out of the business. The United States was on the verge of returning to the pre-vaccine era.

Realizing that American children might soon be denied lifesaving vaccines, the federal government stepped in. On October 18, 1986, the last day of the Ninety-Ninth Congress, legislators passed a bill that protected vaccine makers: the National Childhood Vaccine Injury Act. One month later, President Ronald Reagan signed it into law. The act contained the Vaccine Injury Compensation Program (VICP), which included a list of compensable injuries possibly caused by vaccines. Designed to make things easier for parents, the act specified awards for loss of earnings, lawyers' fees, and up to $250,000 for pain and suffering. At the center of the program stood the injury that had led to the act's passage: seizures and brain damage allegedly caused by pertussis vaccine.

The purpose of the National Childhood Vaccine Injury Act was to allow children to be compensated for vaccine damages without having to go through the expensive process of suing in state courts; to protect pharmaceutical companies from litigation; and to encourage vaccine makers to continue to research and produce new vaccines. The government had taken the burden of litigation off the backs of vaccine makers and put it on its own.

Although legislators had designed the program to satisfy everyone, no one was satisfied. Edward Brandt, from the Department of Health and Human Services, said, "The bill establishes a strong presumption that vaccine is responsible for essentially any adverse condition that happens after immunization unless there is incontrovertible evidence

Henry Waxman proposed legislation that saved vaccines for American children.
(Courtesy of Bloomberg via Getty Images.)

of other causation. This presumption of guilt would undermine public confidence in immunizations." The American Medical Association wanted a panel of scientific experts to determine which vaccine side effects would be compensated, concerned that the task would otherwise fall to members of Congress. And parents were worried that vaccine makers, now largely protected from litigation, would have little interest in making vaccines safer. Henry Waxman, the California Democrat who had sponsored the legislation in the House of Representatives, said, "I recognize the bill I have introduced is probably not the first choice of most parties to this controversy. Manufacturers would undoubtedly prefer greater insulation from liability. Parents of injured children would certainly prefer larger compensation and fewer restrictions on court activity. The Reagan administration would, I am sure, prefer legislation that spends no money."

Despite everyone's misgivings, the National Childhood Vaccine Injury Act saved vaccines. In 1986, the year of the bill's passage, lawyers filed 255 lawsuits against DTP-vaccine makers; by 1996, ten years after the act was passed, they filed only six. Also, the act provided a mechanism to inform parents about vaccine safety, a system to independently review vaccines, and the means to report suspected side effects through the Vaccine Adverse Events Reporting System, or VAERS.

In May 1982, Kathi Williams and Dissatisfied Parents Together had stood before Paula Hawkins's congressional committee and read a list of demands. Only four years later, they'd gotten much of what they'd wanted.

Like their counterparts in the United States, British health officials would also be forced to confront lawsuits, angry parent groups, and distrustful media. But unlike what had happened in the United States, where vaccines were almost eliminated, the outcome in England would be different. The controversy would lead to one of the most unusual and dramatic product liability cases in modern history, and end with a surprising answer to the question of whether pertussis vaccine had caused permanent harm at all.

CHAPTER 3

A Crude Brew

Nothing in life is to be feared—only to be understood.
—MARIE CURIE

At the beginning of *DPT: Vaccine Roulette*, Lea Thompson asked Gordon Stewart to describe the pertussis vaccine. Stewart said it was "a crude brew of those bacteria and all their growth products." Stewart's description was an understatement.

Bordetella pertussis was first grown in a nutrient-rich broth in 1906. In the 1930s, Pearl Kendrick and Grace Eldering made a vaccine by simply killing pertussis bacteria with carbolic acid, an antiseptic. In 1939, they tested it. Kendrick and Eldering studied more than 4,000 children, giving their vaccine to half; during the next four years, they watched to see who got sick and who didn't. The results were clear: whereas 348 unvaccinated children got whooping cough, only 52 vaccinated children suffered the disease. Ten years later, in 1948, the pertussis vaccine was combined with diphtheria and tetanus vaccines to make DTP. Although diphtheria, tetanus, and pertussis vaccines were given in the same syringe, the vaccines were quite different. That's because scientists had a much better understanding of how diphtheria and tetanus caused disease.

Like pertussis, diphtheria is caused by a bacterium: *Corynebacterium diphtheriae*. The bacterium causes a large, painful membrane to form on the back of the throat that can suffocate its victim; it also makes diphtheria toxin, which harms the brain, heart, and kidneys. (In the early 1900s, diphtheria was one of the biggest killers of young children.) Protection against diphtheria is afforded by immunity to this single toxin. So, when researchers wanted to make a diphtheria vaccine, all they had to do was grow toxin-producing bacteria in nutrient fluid, filter the bacteria out of the fluid, and leave the toxin behind. Then they inactivated the toxin with chemicals. Inactivated toxin is called toxoid.

Tetanus vaccine is made exactly the same way. Tetanus, or lockjaw, is caused by the bacterium *Clostridium tetani*. As with diphtheria, tetanus bacteria make just one harmful toxin—and protection against tetanus is afforded by immunity to that toxin. Accordingly, diphtheria and tetanus vaccines each contain only a single protein.

Making a pertussis vaccine, on the other hand, hasn't been easy. That's because the pertussis bacterium doesn't make just one protein that causes disease. It makes many; at least nine pertussis proteins play an important role in infection. Some of these proteins are part of the structure of the bacteria; other proteins, like diphtheria and tetanus toxins, are secreted by bacteria. When Kendrick and Eldering were making their vaccine, they didn't know how many pertussis proteins caused disease. So they took bacteria, grew them in nutrient fluid, and treated the whole concoction with carbolic acid. Their vaccine, made using whole, dead pertussis bacteria, contained more than three thousand pertussis proteins.

When Kathi Williams, Jeff Schwartz, and Barbara Loe Fisher organized Dissatisfied Parents Together, the process of making pertussis vaccine wasn't much different from that used by Kendrick and Eldering forty years earlier. Because the vaccine was so crudely made, it had a higher rate of side effects than any other vaccine. To put this in perspective, in 1982, when Lea Thompson galvanized the public with *Vaccine Roulette*, in addition to the DTP vaccine, children received the combination measles-mumps-rubella (MMR) vaccine and the oral polio vaccine. Measles vaccine contains ten viral proteins; mumps, nine; rubella, five; and polio, fifteen. This meant that the total number of immunological challenges in the measles, mumps, rubella, polio, diphtheria, and tetanus vaccines combined was forty-one, about a hundredth the number contained in the pertussis vaccine alone.

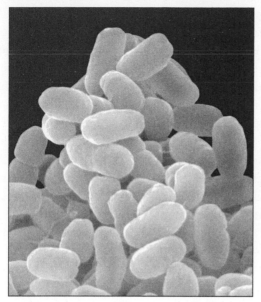

Pertussis vaccine was
the only vaccine given
to American children
made from whole, dead
bacteria. (Courtesy of
Dennis Kunkel
Microscopy/Corbis.)

By the early 1980s, only one study in the United States had carefully examined the side effects of pertussis vaccine. During her program, Lea Thompson introduced the researcher who did it: Dr.
Larry Baraff of the UCLA Medical Center. Baraff explained why he
had done the study: "Because the Food and Drug Administration
was concerned that this sort of public panic might spread [from
England] to the United States, they wanted to document that the
vaccine was safe and not associated with severe consequences."

Baraff's findings were striking. Of every thousand children given
pertussis vaccine, eighty suffered redness and swelling at the site of
injection (more than an inch wide); about five hundred had pain; five
hundred had fever; three had fever greater than 105 degrees; three
hundred felt drowsy; five hundred were fretful; twenty didn't want to
eat; ten cried for more than three hours (and as long as twenty-one
hours); and one had an unusual, high-pitched cry (Kathi Williams's
child suffered this side effect). Further, of every ten thousand children
vaccinated, six suffered seizures with fever and six had decreased
muscle tone and responsiveness that lasted for a few hours. (This side

effect, called Hypotonic-Hyporesponsive Syndrome, can cause children to be pale and limp for hours—devastating for any parent to
watch. Barbara Loe Fisher's child most likely suffered this problem.)
Baraff explained that these side effects were the result of the archaic
technology used to make the vaccine. "I don't think that this is the
type of vaccine that would be produced today," he said. "If this vaccine were produced in 1980, instead of in the 1930s and '40s, there
would be a different type of technology available and we would make
a more purified vaccine." Baraff was right. Taking advantage of advances in protein chemistry and protein purification, by the mid-
1990s, safer pertussis vaccines—containing only two to five pertussis
proteins instead of three thousand—were licensed.

Although transient side effects following pertussis vaccine were
common, that wasn't at issue. The important question raised by
Vaccine Roulette was whether the vaccine could cause permanent
harm, such as epilepsy and mental retardation. Answering this question isn't as easy as it seems. That's because every year in the United
States, in England, and throughout the world, children suffer
epilepsy and mental retardation; this has been true for centuries,
well before the pertussis vaccine was invented. Also, symptoms of
epilepsy and retardation often occur in the first year of life, the same
time that children are receiving three doses of vaccine. Given the
widespread use of pertussis vaccine, most children destined to develop seizures or mental retardation anyway would likely have received it, some within the previous twenty-four or forty-eight hours.
So, the only way to figure out whether the vaccine was the problem
was to study thousands of children who did or didn't get it. If the
vaccine were responsible, the risk of epilepsy or retardation would
be greater in the vaccinated group. At the time of *Vaccine Roulette*,
only one large-scale study of children had been completed: David
Miller's. The next fifteen years, during which other investigators examined this question, were not kind to David Miller's study.

The first flaw with the notion that pertussis vaccine caused brain damage was that it didn't make biological sense. Anyone working in a
hospital knew that natural pertussis infection could cause brain dam-

age because of decreased oxygen in the bloodstream caused by unrelenting coughing. But the pertussis vaccine—made of killed bacteria that didn't grow in the lungs or windpipe—didn't cause coughing.

So what could have caused brain damage? One prevalent theory was that the pertussis vaccine contained small amounts of endotoxin: part of the surface of a variety of bacteria (including pertussis) and a very potent poison. In 1978, a researcher named Mark Geier published a paper claiming that commercial preparations of pertussis vaccine contained small quantities of endotoxin. Because even a little endotoxin can have devastating effects, Geier reasoned that severe reactions to pertussis vaccine might be caused by endotoxin. The problem with this logic is that endotoxin causes brain damage by inducing a cascade of events that includes fever, rapid heart rate, chills, low blood pressure, shock, and decreased oxygen to the brain. Indeed, the one symptom that occurs in everyone experimentally inoculated with endotoxin is fever. But many children with seizures and retardation following pertussis vaccine never had fever. And pertussis vaccine didn't cause low blood pressure or shock, another common response to endotoxin.

Epidemiological studies didn't support Miller's study, either.

In 1956, the Medical Research Council in England studied more than thirty thousand children for two years. It couldn't find even one who had suffered brain damage as a result of the pertussis vaccine.

In 1962, Bo Hellström at the Karolinska Institute in Stockholm studied eighty-four healthy infants who had received DTP and later suffered high fever or decreased responsiveness. Hellström performed electroencephalograms on children six and twenty-four hours after vaccination, reasoning that if the vaccine was affecting the brain, then the EEGs, which can detect even slight alterations in brain wave activity, should be abnormal. But they weren't. All the children had perfectly normal EEGs.

In 1983, two years after David Miller's study was published, T. M. Pollack and Jean Morris, researchers at the Public Health Laboratory and Department of Community Medicine in London, published their own study. They analyzed 134,700 children in the North West Thames Region of England who had received three doses of DTP and compared them to 133,500 children who had received DT alone. Their study enabled them to isolate the effect of the pertussis component of the vaccine. They, too, couldn't find what David Miller had found.

Also in 1983, three neuropathologists in England studied the brains of twenty-nine children whose deaths had been blamed on pertussis vaccine. Some had died within a week of getting the vaccine; others had suffered epilepsy, mental retardation, or physical disabilities. The investigators were looking for anything that tied these cases together—some indication that the vaccine had caused the problem. But no distinct pathological finding linked the cases.

In 1988, researchers in Denmark took advantage of a natural experiment. Before April 1970 children in Denmark were vaccinated with the DTP vaccine at five, six, seven, and eighteen months of age. But after April 1970 they were given the pertussis vaccine at five and nine weeks and again at ten months of age. Investigators reasoned that if epilepsy were a consequence of receiving pertussis vaccine, then the onset of seizures should change with the changing schedule. But it didn't.

Six years after the airing of *Vaccine Roulette*, a study was finally performed on American children. In 1988, researchers from the department of epidemiology at Harvard's School of Public Health and the Group Health Cooperative of Puget Sound in Seattle examined the records of more than thirty-five thousand children. They wanted to determine whether epilepsy was more common in those recently vaccinated. It wasn't.

Two years later, Marie Griffin and colleagues from Vanderbilt University published a study of more than thirty-eight thousand children in Tennessee, looking for a relationship between DTP and brain damage. Again, they were unable to find what David Miller had found.

The number, reproducibility, and consistency of these studies prompted a response from public health agencies. In 1989, the British Pediatric Association and the Canadian National Advisory Committee on Immunization concluded that pertussis vaccine had not been proved to cause permanent harm.

In 1990, the same year Marie Griffin published her paper, Gerald Golden—Shainberg professor of pediatrics, director of the Boling Center for Developmental Disabilities, and professor of neurology at the University of Tennessee—reviewed the evidence. Golden was unequivocal in his conclusions. "A syndrome of pertussis vaccine encephalopathy was first reported fifty-six years ago," he wrote, referring to the 1933 report of two deaths following pertussis vaccine in Copenhagen. "Analysis of the recent literature, however, does not support the existence of such a syndrome and suggests that *neurological events after immunization are chance temporal associations*." James Cherry, the pediatric infectious diseases specialist from UCLA, seconded Golden's conclusions regarding the vaccine–brain damage link, writing that it was "time to recognize it as the myth that it is."

In 1991, the Institute of Medicine—an independent research institute within the U.S. National Academies of Science—concluded that the association between pertussis vaccine and brain damage remained unproved. An ad hoc committee for the Child Neurology Society agreed, stating, "Case reports have raised the question as to whether there is an association between pertussis vaccine and progressive or chronic neurological disorders, but controlled studies have failed to prove an association."

More evidence mounted.

In 1994, researchers from the University of Washington and the Centers for Disease Control and Prevention teamed up to do yet another study. They evaluated more than two hundred thousand children in Washington and Oregon and concluded: "This study did not find any statistically significant increased risk of onset of serious acute neurological illness in the seven days after DTP vaccine."

On March 10, 1995, seizure disorder following DTP vaccine was removed from the Vaccine Injury Compensation Program's list of compensable injuries because "no medical evidence" existed to support the presumption—ironic, given that the program was born of this specific concern.

Finally, in 2001, researchers working with a group of health-maintenance organizations performed the clearest, most definitive study to date. Using computerized records, investigators analyzed the occurrence of seizures in three hundred and forty thousand children given DTP compared with two hundred thousand children who received no vaccine. They concluded: "There are significantly elevated risks of febrile seizures after receipt of DTP vaccine, but these risks do not appear to be associated with any long-term, adverse consequences." (Although frightening to witness, seizures caused by fever, which occur in as many as 5 percent of young children, don't cause permanent harm.)

No one had been able to repeat David Miller's study—remarkable, given its impact.

Miller and his team had worked hard. They had evaluated data from 1,182 children accumulated from July 1, 1976, to June 30, 1979. Further, their study wasn't limited to London or England; they had also evaluated all children with neurological illnesses in Scotland and Wales. Miller and his team had assessed reactions to more than two million doses of pertussis vaccine; and they'd spent millions of dollars doing it. No study of the vaccine had ever been so thorough or so large. So why weren't any subsequent investigators able to find what David Miller had found? The answer wouldn't come from a scientific laboratory, or an academic institution, or an independent research organization. It would come, surprisingly, from a British courtroom.

In the United States, hundreds of lawsuits resulted in many trials. In England, on the other hand, the DTP scare resulted in only three trials. The last would show precisely what had gone wrong with David Miller's study.

The first trial took place in Edinburgh, Scotland, in 1985, centering on a little boy named Richard Bonthrone.

Bonthrone had received his first dose of DTP vaccine when he was four months old. Three months later, he received his second. Nine days after that, he had the first of many seizures. At the time of the trial, Richard was nine years old with the mental age of a six-month-old. He ate only liquid food and couldn't do anything for himself.

The judge, Lord Jauncey, wrote that Richard's "only enjoyment of life appears to be in recognition of his mother's voice and in travel by motor car." Richard's parents, John and Iris, sued their doctor, the Department of Health, and a visiting nurse for £145,000.

Two experts testified at Bonthrone's trial. Dr. John Stephenson, a pediatric neurologist from the Hospital for Sick Children in Glasgow, said that while he accepted that DTP precipitated seizures following fever, he "was unconvinced that permanent brain damage resulted." Next up was David Miller, who said, "The risk of developing encephalopathy nine days after being vaccinated [is] so small as to be incapable of statistical measurement." Judge Jauncey ruled in favor of the defense. But the Bonthrone trial hadn't refuted Miller's study. Miller had never claimed that DTP caused harm nine days after a dose, only within three days of a dose.

Another trial quickly followed.

Johnnie Kinnear was fourteen months old when he received a dose of pertussis vaccine. According to his mother, Johnnie had a seizure seven hours later. The next morning, she took her son to the doctor who dismissed her concerns, saying it was "normal for children to have reactions" and there was "nothing to worry about." But the seizures continued, every day for many months. The Kinnears sued the Department of Health, their doctor, the North West Thames Health Authority, and the Wellcome Foundation, which manufactured pertussis vaccine in England. The legal aid board intervened, allowing the suit to proceed only against the health authority and the doctor. The Wellcome Foundation was excused in part because it was one of several manufacturers of pertussis vaccine in England (Glaxo and Lister also made the vaccine); and it was unclear which brand Kinnear had received. But, in an unprecedented move, Wellcome decided to enter the case anyway—in essence, choosing to be sued. Wellcome was tired of the negative press surrounding pertussis vaccine and wanted to settle the issue once and for all. It, too, wanted its day in court. This was the first time that a drug maker in England had defended the safety of one of its products in a full court hearing.

The Kinnear trial began on March 17, 1986, and included many experts in the fields of virology, epidemiology, and statistics; among them were John Wilson, whose paper had ignited fears about the vaccine in England; Gordon Stewart, who had been prominently featured in *Vaccine Roulette*; and David Miller. The judge in the trial was Murray Stuart-Smith. Fifty-seven years old and the father of six, Stuart-Smith had graduated with first-class honors from Cambridge University. He would soon become a celebrated jurist and acquire the title of "Sir." (Stuart-Smith would later sort out the tragedy that occurred at Hillsborough Stadium on April 15, 1989, when ninety-six Liverpool football fans were killed in a melee.)

The lawyer for the Kinnears was Julian Priest, who began by taking his star witness, Gordon Stewart, through early case reports. Stewart talked about the 1933 Copenhagen report, which described two Danish children who had suddenly died after receiving vaccines. He talked about the study from Harvard Medical School by Byers and Moll, who had reported fifteen children with irreparable damage following pertussis vaccine. He talked about Justus Ström's paper from Sweden claiming that the vaccine caused permanent damage in one in thirty-six hundred children, far greater than the one in one hundred thousand later claimed by David Miller. Stewart concluded, "No matter how scrupulously the vaccine was prepared, one could not avoid including toxic products."

The Kinnear case had two large holes. The first problem was a discrepancy between the mother's testimony and the medical record. Kinnear's mother testified that her son's seizures began seven hours after he received DTP. But the medical record showed that Johnnie Kinnear's first seizure had occurred five months, not seven hours, later. The contradiction was undeniable, causing Judge Stuart-Smith to say, "The mother's evidence was crucial as to the date of onset of symptoms, which she said occurred very shortly after the vaccination. Unfortunately, she was not telling the truth. And this was apparent to all, including the plaintiff's advisors."

The second problem was Gordon Stewart. No one's reputation would suffer more from the Kinnear trial than his. During the trial,

Stewart had repeatedly misquoted details of David Miller's study. But his greatest embarrassment came during his discussion of another study. Stewart had stated: "Levine and Wenk described a hyperacute allergic encephalomyelitis which occurred in children who appeared to have been sensitized by a previous dose of pertussis vaccine." Anthony Machin, the Wellcome Foundation's attorney, asked Stewart, "Do you remember anything about the age of the children?" "No, not offhand," replied Stewart. Machin pressed on: "Or the ethnic origin?" "No, I cannot remember that," said Stewart. "It was an American study, I know that." Machin then handed the study to Stewart, who looked through the paper and then up at Machin, embarrassed. Stewart apologized for his mistake. The study hadn't been done on children; it had been done on rats. The gaffe marked the end of Gordon Stewart's role in damning pertussis vaccine. During a subsequent trial, he was labeled "an evidential liability." Although the Kinnear trial took twenty-nine days, included many witnesses, and cost the government £1 million to litigate, it ended without a verdict.

The litigation, however, didn't end.

The trial that ended all pertussis vaccine trials in England began on February 1, 1988. Again the judge was Lord Justice Murray Stuart-Smith. The plaintiff was Susan Loveday, a mentally disabled seventeen-year-old who lived in Bradford-upon-Avon in Wilshire. Susan's story was similar to the others. In 1971, George Renton, Susan's doctor, administered the first dose of DTP after which Susan "had a high temperature and local inflammation and was sleepy but crying a great deal." Her mother noticed that she wasn't as lively and that "one eye seemed odd." One year later, Renton gave Susan her second dose of DTP. This time she "had a similar reaction and screamed all night." One month later, she was referred to another pediatrician who noticed that she looked "rather odd and hypotonic [floppy]." Then Renton gave her a third shot of DTP. By the time she had her day in court, Susan Loveday was severely retarded.

Loveday was the lead case in a class-action lawsuit that included two hundred other children with similar stories. Most had been gathered by Rosemary Fox, the head of Britain's Association of Parents of Vaccine-Damaged Children. Unfortunately, as had happened with Johnnie Kinnear, under closer inspection Susan Loveday's story started to unravel. It shouldn't have come as a surprise. In fact, when Rosemary Fox had sent out questionnaires to find vaccine-damaged children, she had weeded out Loveday's story as improbable. Even Gordon Stewart, an unwavering advocate for the cause, said, "She was not vaccine-damaged. She was damaged before."

Stuart-Smith had learned his lesson from the first trial. Rather than simply examining the merits of Loveday's claim to see if it was a true case of vaccine damage, he decided to divide the trial into two parts. The first would be devoted to determining whether pertussis vaccine could cause permanent harm. If the answer was yes, the second part would determine who qualified for compensation. As in the Kinnear trial, the Wellcome Foundation voluntarily joined in. Wellcome wanted to see the data on which David Miller had based his conclusions—to see the stories of the children who were allegedly harmed by pertussis vaccine.

The case of *Loveday v. Renton and Wellcome Foundation Ltd.* took four months, involved nineteen expert witnesses, and cost more than £1 million to conduct. The judgment was more than one hundred thousand words long and included fourteen chapters and six appendices. The verdict was so long that it took Stuart-Smith two full days to read it; and so detailed that it left little room for appeal. Although the Loveday trial considered a great many issues, it was a trial of one study and one study only: David Miller's. Under the bright lights of Justice Stuart-Smith's courtroom, Miller's study fell apart.

Stuart-Smith began the trial by expressing sympathy for parents who believed their children had been damaged by DTP. "For many of [the parents] the answer must seem obvious," he said. "Their child is vaccinated and within a relatively short time thereafter, perhaps hours or days, [he] becomes seriously ill. [He] may have convulsions, become unconscious, show signs of paralysis; there may be prolonged screaming or vomiting. Later, there are signs of permanent mental handicap; blindness, deafness, paralysis or motor impairment, epilepsy, severe mental retardation. The one event must have caused the other, especially when the time interval between vaccination and onset of symptoms is measured in hours rather than days." But Stuart-Smith wasn't going to be swayed by temporal associations. Quoting Samuel Johnson, he said, "It is incident I am afraid, in physicians above all men, to mistake subsequences for consequences." Then he elaborated. "Where given effects, such as serious neurological disease or permanent brain damage, occur with or without pertussis vaccination, it is only possible to assess whether the vaccine is a cause, or more precisely a risk factor, *when the background incidence of the disease is taken into account.* The question therefore is, does the effect occur more often after pertussis vaccination than could be expected by chance?"

By quoting Samuel Johnson, Stuart-Smith had hit on the central issue in the trial: the power of anecdote. For the parents of Richard Bonthrone, Johnnie Kinnear, and Susan Loveday the issue was clear. Their children had been fine until they'd received the pertussis vaccine. But the mere fact that one event follows another doesn't necessarily mean that it caused the other. (Unfortunately, it's hard to fight anecdote with statistical data. An emeritus professor at Duke University School of Medicine tells the story of a friend who took his four-month-old son to get a DTP vaccine. He waited and waited in line, eventually tiring and going home, his child never having received the vaccine. Several hours later the father went to wake the child, only to find that he had died, presumably from Sudden Infant Death Syndrome. One can only imagine what the father would have felt if his son had received DTP several hours earlier. Presumably, no study would have convinced him of anything other than that the vaccine had killed his son.)

Stuart-Smith began by singling out the man who had started it all: John Wilson. After acknowledging that Wilson's paper had been circumspect and "tentative in its conclusions," the judge was angry that he had included two children in his report who had never received pertussis vaccine. "Dr. Wilson is entirely convinced that the vaccine on rare occasions causes brain damage," said Stuart-Smith. "He formed this view at a very early stage and he is completely committed to it. An example of this, small in itself but perhaps revealing, is to be found in the two DT cases in [his paper]. When he gave his evidence Dr. Wilson had forgotten that he had previously known that these were DT and not DTP cases, and so he stated at first that they were DTP. I was not convinced by his explanation that the matter was of no consequence and did not affect the thrust of the article. I think he had forgotten it because it was inconvenient." Chastened, John Wilson would never again carry high the torch that pertussis vaccine caused brain damage.

Then Stuart-Smith turned his attention to David Miller. Miller was certainly aware of the furor surrounding his study; aware that Rosemary Fox had formed the Association of Parents of Vaccine-Damaged Children; aware that politicians had rallied in support of these children; aware that several doctors, such as Gordon Stewart, had become vocal advocates for the cause; and aware that the press and public knew full well that the national government had funded his study. Miller didn't want to appear to have whitewashed the issue. So he bent over backward to report any possible problems with the vaccine. Earlier in the trial, one of Miller's collaborators had stated, "Because it was essential that pertussis vaccine should not appear from this study safer than it actually is, steps were taken at every stage to ensure that the results would overestimate rather than underestimate the risks." An example of this type of bias appeared in instructions to physicians. "If there is doubt, code the worst picture," they read. David Miller described what it was like to do a study under the bright lights of the press and public. "We were extremely anxious," he said. "One has to remember the context in which the study was done, which was that there was a great deal of alarm about the possibility of there being damage associated with the vaccine. . . . It was extremely important that our results were not subject to criticism by those who might say that we were endeavoring to avoid demonstrating the possibility of a neurological damage associated with a vaccine. . . . It was very important that we avoided that kind of accusation." Stuart-Smith was unsympathetic: "I think it can be said that this demonstrates a conscious over-anxiety to appease what I may call the vaccine-damage lobby, which may have led to decisions being biased against the vaccine."

There were bigger problems. Stuart-Smith was unhappy that Miller's study, due to political pressure, was published prematurely. "Unfortunately, owing to what Professor Miller described as political though not party-political pressure, the report of the [study] had to be published before the data were complete. As a result, it deals only with the first one thousand of the total of 1,182 cases. More importantly, the data relating to the final outcome were incomplete." When the data were complete, when the full follow-up was concluded, and when the details of all of the cases were finally revealed, a very different picture emerged.

In his study David Miller maintained that seven children had developed brain damage within a week of receiving DTP. But a closer look showed that these cases weren't what they were claimed to be. Three of the children had been incorrectly labeled as brain damaged when in fact they were normal both before and after vaccination. Three others had suffered viral infections and one Reye's Syndrome (a severe neurological problem found later to be caused by aspirin, not vaccines). Miller's conclusions lay in ruins.

In March 1988, Stuart-Smith delivered his verdict: "On all the evidence, a plaintiff had failed to establish, on a balance of probability, that pertussis vaccine used in the United Kingdom and administered intramuscularly in normal doses could cause permanent brain damage in young children." David Salisbury, currently director of immunization for England's Department of Health, remembered the impact of Stuart-Smith's verdict: "I thought that for a judge who was not a trained epidemiologist it was an extraordinarily insightful analysis of all of the available evidence. The [verdict] clearly gave us much more authority to be clear about the risks of pertussis and the risks of not being vaccinated."

No other trials against pertussis vaccine would ever appear in a British courtroom. Soon after Judge Stuart-Smith concluded in England that pertussis vaccine didn't cause permanent brain damage, Judge John Osler made an identical decision in a landmark case in

Canada. Osler concluded that while the "post-pertussis theory was popular some years ago," further study and technology refuted the contention.

Stuart-Smith's trial had gone a long way toward putting to rest the notion that pertussis vaccine caused epilepsy and mental retardation. But it never addressed the real cause or causes of the problem. That's because science hadn't advanced far enough to answer the question of what caused seizures. During the next two decades, that would change. And the answer wouldn't come from England, where the notion that pertussis vaccine caused brain damage had first gained international attention, or from the United States, where vaccines had almost been eliminated by lawsuits. Rather, it would come from a relatively unknown researcher on the east coast of Australia.

Seizures are more common than most people realize, typically occurring in the first year of life. Indeed, every year as many as one hundred and fifty thousand children in the United States develop seizures caused by fever. Most of these children never have seizures again. However, every year about thirty thousand children, whose first seizure may or may not have been associated with fever, develop epilepsy. When Lea Thompson was putting the pertussis vaccine on trial in the media, or Paula Hawkins in Congress, or Stuart-Smith in a British courtroom, seizure disorders in infants were not very well understood. But, during the next twenty years, neurologists made tremendous strides in sorting out different seizure types based on clinical symptoms, EEG patterns, age of onset, and response to therapies. More important, with the availability of genetic probes, the specific genes that caused many of these epilepsy syndromes were found. As of 2009, the genetics of at least fifteen different epilepsy syndromes had been identified.

It was with this knowledge—and these genetic tools in hand—that Samuel Berkovic, director of the Epilepsy Research Center, scientific director of the Brain Research Institute, and the Laureate Professor in the department of medicine at the University of Melbourne, decided to revisit the question of what was causing epilepsy and mental retardation in children after DTP vaccine. Berkovic was

Samuel Berkovic, a neurologist at the University of Melbourne, was the first to figure out what had caused seizures and mental retardation in the children featured in Lea Thompson's *DPT: Vaccine Roulette*. (Courtesy of Dr. Samuel Berkovic.)

particularly interested in a type of epilepsy called Dravet's Syndrome: a genetic disorder. First described by Charlotte Dravet in 1978, the syndrome included children who had seizures and mental retardation exactly like those described in Lea Thompson's program. But Charlotte Dravet had described her syndrome in children, not adults. And Berkovic only took care of adults. "I just vividly remember seeing in the clinic this woman who was in her mid-forties," he recalled. "I had treated her for fifteen or twenty years. [She had] a very devoted mother who would bring her to see me. And it's actually hard to recognize [Dravet's Syndrome] in adults because the clinical features have their evolution in the first couple years of life. Then the penny dropped. This is Dravet's Syndrome!" Samuel Berkovic was the first person to postulate that adults with seizures and retardation who were labeled vaccine-damaged as children might instead be suffering from Dravet's. He remembered the emotional burden that the mother had carried for more than forty years: "Why did she get her daughter vaccinated? If only she hadn't gotten her vaccinated this terrible thing wouldn't have happened. And she carried the guilt and grief of all of this. So we started looking aggressively to collect the cases."

In 2006, Samuel Berkovic evaluated fourteen people with severe epilepsy and mental retardation. All had developed their first seizures between two and eleven months of age; all had received the pertussis vaccine within the previous forty-eight hours; and several had been compensated for damage allegedly caused by the vaccine. Some had had fever following the vaccine, but most hadn't. Berkovic believed that all suffered from Dravet's Syndrome. So he looked to see whether they had a genetic defect that caused the problem. He found that eleven of the fourteen had a defect in the gene that regulates the transport of sodium in brain cells. (The specific gene is called *SCN1A*; and the specific disorder, a sodium channel transport defect.) Recognizing that vaccines can't change a child's genetic makeup—and that 100 percent of children with *SCN1A* defects will have seizures and mental retardation independent of whether they receive vaccines—Berkovic wrote, "The identification of a genetic cause of encephalopathy in a particular child should finally put to rest the case for vaccination being the primary cause." In the last few sentences of his landmark paper, Samuel Berkovic made a plea on behalf of children with epilepsy: "Correct diagnosis will reassure the family as to the true cause, remove the blame of having vaccinated the child, direct appropriate treatment, and allow realistic planning for prognosis."

Berkovic was surprised by the reaction to his paper. "I was very excited about this [work]," he recalled. "I thought it was one of the most, if not *the* most, important thing I'd ever done." Following publication, several commentaries praised the work, one calling it the most outstanding paper of 2006. "Most of my neurologist colleagues thought that the paper was really important because they're the people that see these cases directly. But among the vaccine community I couldn't get any traction. We were trying to get a much bigger study together and collect more cases and we just couldn't. I got a bit dispirited and gave up. Why didn't it get traction? I don't know. It befuddled me. Maybe it's because it was something from a bygone era."

After Berkovic's paper, it was clear that all the time spent by parents to get health officials to admit that pertussis vaccine had permanently harmed children, all the money spent by pharmaceutical

companies to compensate alleged victims, all the work of lawmak-
ers to create a system to deflect lawsuits away from these companies, and all the ink devoted by the media to support these children
and their parents had been an enormous diversion from the real
cause of the problem.

Although parents of children with epilepsy and mental retardation were wrong in believing that pertussis vaccine was at fault, they
were right in several respects: First: it's likely that their children had
appeared perfectly normal before they received the vaccine. Many
infants with epilepsy and retardation syndromes have no symptoms
for the first few months of life.

Second: many doctors had assumed that seizures following DTP
had been febrile seizures and falsely reassured parents that these
were common and nothing to worry about. During Paula Hawkins's
hearings into the safety of the pertussis vaccine, Marge Grant,
whose son Scott had been prominently featured in *Vaccine Roulette*,
submitted written testimony. "Before I go into the grave injustice
we've encountered through the entire court system, including the
U.S. Supreme Court," she wrote, "I wish to re-emphasize that Scot-
tie did NOT display a 'fever' after any of the shots." Consistent with
Grant's testimony, most of the children in Berkovic's study hadn't
had seizures that were precipitated by fever, either.

Third: parents were right in pointing out that no one else in the
family had had epilepsy. During *Vaccine Roulette*, Marge Grant
took aim at the Food and Drug Administration (FDA), elicited the
support of Jimmy and Rosalynn Carter, and skewered pharmaceutical companies for misrepresenting, hiding, or shredding data. But
during Hawkins's hearing, Grant also directed her anger toward
doctors. "As I look back at this nightmare and the severity of Scott's
injury," she recalled, "I do not believe he is just one unusual, obscure case, who displayed almost no early symptoms, yet suffered
profound permanent damage. I am convinced there is an extremely
fine line, such as in Scott's instance, where major vaccine problems
exist, but actually go undetected by parents until much later, when
they are finally told the child's developmental deficits must be genetic. Can you imagine the trauma that must hit a parent? If they

felt it was genetic, they would probably never have another child."
Grant knew that neither her own nor her husband's family had a
history of epilepsy or retardation. But she hadn't understood that
not all genetic problems are inherited. Samuel Berkovic subsequently addressed this issue in his study. He noted, "In nine of the
eleven patients with *SCN1A* mutations for whom samples from
both parents were available, the mutations were absent in parental
DNA and thus arose *de novo*." This meant that the genetic mutation had occurred spontaneously during conception.

Soon after Samuel Berkovic published his paper, Simon Shorvon
and Anne Berg, researchers in England and the United States, wrote
an editorial discussing the importance of its findings. "If properly
communicated," they wrote, "the Australian findings should also
do much to improve the public's understanding of the true risks and
the safety of this vaccine." During the pertussis scares in England
and the United States, scares that spilled over into virtually every
developed country in the world, thousands of media outlets falsely
alerted parents to the harm caused by pertussis vaccine. But when
Samuel Berkovic finally found the answer to the question of what
had actually caused the problem, no one noticed. Not a single newspaper, magazine, or radio or television program carried the story.

CHAPTER 4

Roulette Redux

If you watch TV you know less about the world than
if you just drink gin out of a bottle.
—GARRISON KEILLOR

Lea Thompson began *DPT: Vaccine Roulette* in 1982 with a clear
statement of purpose: "Our job in the next hour is to provide
enough information so that there can be an informed discussion
about this important subject. It affects every single family in America." Unfortunately, throughout *Vaccine Roulette*, Thompson did
much to misinform her viewers.

When Thompson interviewed Larry Baraff of the UCLA Medical
Center about his study of the side effects of pertussis vaccine, she
said, "The study estimates one of every seven hundred children had
a convulsion or went into shock." But Baraff's study never mentioned the word *shock*. Rather, it described nine children who developed hypotonic hyporesponsive episodes, in which a child is
temporarily limp and unresponsive. The hallmark of shock is low
blood pressure resulting in decreased blood flow to critical organs.
No such problem was mentioned in Baraff's study nor has such a
problem ever been found to be caused by pertussis vaccine.

Toward the beginning of *Vaccine Roulette*, Thompson stood fac-
ing the camera with a book in her hand: the *Physicians' Desk Reference*, or *PDR*. Published once a year, the *PDR* contains a
compendium of package inserts for medicines and vaccines. "The

Physicians' Desk Reference, prepared by manufacturers, says the 'P' part of the DPT vaccine is a possible link to Sudden Infant Death Syndrome," said Thompson, briefly showing a highlighted section of the book. Thompson should have kept the book on the screen longer, allowing the viewer to read what was actually written. "The occurrence of Sudden Infant Death Syndrome (SIDS) following receipt of DTP vaccine has been reported," it read. "The significance of these reports is unclear. It should be kept in mind that the three preliminary immunizing doses of DTP are usually administered to infants between the ages of 2 and 6 months, and that about 85 percent of SIDS cases occur in the period 1 to 6 months of age." In other words, the association is likely to be coincidental. Indeed, a study published in 1982 showed that pertussis vaccine didn't cause SIDS. Despite this evidence Thompson ended her program by saying, "It is difficult to come up with a definitive answer as to how many children are being severely damaged or are dying from the DPT vaccine."

When Lea Thompson introduced the story of Scott Grant, she implied that infantile spasms, a type of seizure, were caused by DTP. Infantile spasms have a unique pattern of brain wave activity on EEG, so they're easy to diagnose. Because they're easily diagnosed, they're easily studied. Five years before *Vaccine Roulette*, a study from Denmark clearly showed that DTP didn't cause infantile spasms. This study was well known, having been included in review articles and book chapters for several years. Subsequent studies confirmed the Denmark study. Unfortunately, by including Scott Grant's story, Thompson left her viewers with the false impression that his epilepsy was caused by DTP.

Thompson also understated the severity of pertussis infection. While showing many children with alleged brain damage following vaccination, she showed only one child with whooping cough. And she said that the massive pertussis epidemic in England in the late 1970s—during which thousands of children were hospitalized and hundreds died—wasn't as bad as doctors had claimed. Gordon Stewart referred to it as the "so-called epidemic." This created the impression that if Americans stopped giving pertussis vaccine, nothing bad would happen. When the Centers for Disease Control and

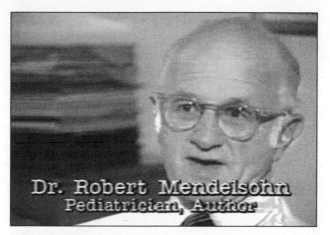

Robert Mendelsohn, a self-described medical heretic, was prominently featured in *DPT: Vaccine Roulette*. (Courtesy of WRC-TV/NBC News.)

Prevention and the American Academy of Pediatrics criticized Thompson for failing to point out the devastating effects of whooping cough, she argued that she couldn't find any evidence that it was a problem in the United States. "In fact, whooping cough disease in this country is almost gone," she said. In 1982, when *Vaccine Roulette* aired, three thousand children were hospitalized with whooping cough and ten died from the disease.

Perhaps most disturbing, Thompson misrepresented published information. "The 1977 *Red Book* lists high fever, collapse, shock-like collapse, inconsolable crying, convulsions, and brain damage as reactions to the DPT vaccine," she said. "Those complications are associated with varying degrees of retardation." However, nowhere in the *Pediatric Red Book* of 1977 is pertussis vaccine listed as a cause of brain damage or retardation. William Foege, who had worked on the successful campaign to eradicate smallpox and was head of the CDC, said of *Vaccine Roulette*, "If journalistic malpractice was a recognized entity, I think this program would qualify."

Thompson chose several experts to support her contention that pertussis vaccine was dangerous. Her choices were unfortunate.

Early in her program, Thompson introduced Dr. Robert Mendelsohn: "author, lecturer and former head of pediatric departments at the University of Illinois Medical School and the Michael Reese Hospital in Chicago." In fact, Robert Mendelsohn had never been the head of pediatric departments at the University of Illinois Medical School or the Michael Reese Hospital.

Mendelsohn had made a career of setting himself against the medical establishment, including writing three books: *Confessions of a Medical Heretic*, *Male Practice: How Doctors Manipulate Women*, and *How to Raise a Healthy Child in Spite of Your Doctor*. During the program, Mendelsohn said of the pertussis vaccine, "It's probably the poorest and most dangerous vaccine that we now have," "the statistics of this country are wrong," and "the dangers are far greater than any doctors have been willing to admit." But it wasn't that Mendelsohn thought that children shouldn't get pertussis vaccine; he believed they shouldn't get any vaccines. And it wasn't only vaccines that he thought were useless; he also opposed water fluoridation, coronary bypass surgery, licensing of nutritionists, and mammography screening for breast cancer.

Mendelsohn had enormous disdain for his profession. In 1979, for example, he wrote, "Doctors turn out to be dishonest, corrupt, unethical, sick, poorly educated, and downright stupid more often than the rest of society. When I meet a doctor, I generally figure I'm meeting a person who is narrow-minded, prejudiced, and fairly incapable of reasoning and deliberation. Few of the doctors I meet prove my prediction wrong." Regarding surgeons, he said in 1983, "There is never enough blood in the hospital temples of Modern Medicine to satisfy the surgeon's desire as he seduces his victims—primarily women—virgin and otherwise—to mount the holy altar so he can carry out his ritual mutilations. The wild blood-lust, starting with animal vivisection and proceeding to human mutilation, stamps Modern Medicine as the most primitive weapon this world has ever seen."

Mendelsohn, whose distrust of modern medicine made him a perfect spokesman for Thompson's program, loved *Vaccine Roulette*, saying it was "the greatest thing since apple pie. For the first time the American people got the truth about pertussis vaccine."

Thompson also interviewed a microbiologist named Bobby Young, asking him to comment on an apparent government conspiracy to hide the truth. "I was employed at the Bureau of Biologics [part of the Food and Drug Administration] for several years," said Young, "and it is my opinion that they very much do not wish to know adverse reactions."

"Why?" asked Thompson.

"Well, this will complicate their lives considerably," replied Young.

Then Thompson asked Young about Larry Baraff's study. "The UCLA study found more reactions than had ever been seen before," she said. "The study estimates that one of every thirteen children had persistent or high-pitched crying after the shot."

Young replied, "This may be indicative of brain damage in the recipient child."

There were a few problems with this exchange. First: Thompson implied that Young had worked on the pertussis vaccine while at the FDA, but he hadn't. Second: Bobby Young had no specific expertise in neurology or pediatrics and had never taken care of a patient, so he was ill-equipped to comment on the relationship between crying and brain damage. It's not only that crying—high-pitched or otherwise—doesn't cause permanent harm; even seizures, whether associated with fever or not, do not of themselves cause brain damage. Had Bobby Young been a clinician, he likely would have known that. Finally, and most disturbing, Young appeared to be answering questions about Larry Baraff's study—impossible, given that Young was interviewed and had died before the study was published. This raised questions about whether Thompson had juxtaposed questions and answers. Thompson denied the allegation. "I am not going to risk my reputation by moving someone's words around," she said. "That speaks to my journalistic integrity."

But Young's interview wasn't the only one that raised questions. Ed Mortimer, a former chairman of the AAP's committee on infectious diseases, said that during his interview Thompson asked the same question "repeatedly in slightly different ways, apparently to develop or obtain an answer that fit with the general tone of the program." Mortimer felt that "cutting and splicing remarks taken out of context gave a very different meaning from what I intended or what I believe."

Gordon Stewart was another of Thompson's experts with questionable credentials.

At the beginning of *Vaccine Roulette*, Thompson introduced Stewart as a member of the United Kingdom's Committee on the Safety of Medicines. But Stewart had never been a member of that group. Stewart said, "I believe that the risk of damage from the vaccine is now greater than the risk of damage from the disease." What Thompson didn't say was that five years earlier, in 1977, Stewart had published an article claiming not only that the pertussis vaccine was unsafe but that it didn't work. Like Mendelsohn, Stewart believed the decline in pertussis in the United Kingdom had nothing to do with the vaccine; it was simply a matter of improved sanitation. Given abundant evidence that the incidence of pertussis is inversely related to vaccine use, Stewart's proclamations were at best ill-informed and at worst dangerous. Indeed, in 1977, after British health officials called for a pertussis immunization campaign, Gordon Stewart cried foul. "I accuse the committee [on the safety of medicines] of deceit," he proclaimed. "There are no grounds for saying a major epidemic is on the way and I don't agree with the way their figures have been collected." During the next two years, more than a hundred thousand children were hospitalized and six hundred killed by pertussis.

Soon after *Vaccine Roulette*, Gordon Stewart retired. But he didn't retire from educating the press and the public about infectious diseases. In 1981, one year before Thompson's program, an unusual organism then called *Pneumocystis carinii* killed five homosexual men in Los Angeles; all of these men had severe immunological deficiencies. The CDC eventually called it Acquired Immune Deficiency Syndrome, or AIDS. At first, no one knew what caused the disease. But by 1983, a group of French researchers headed by Luc Montagnier had found the culprit: a virus later named human immunodeficiency virus (HIV). (Montagnier won a Nobel Prize for his discovery.)

Stewart didn't buy it. He took on a medical profession that he believed had been duped into believing that HIV caused AIDS with the same ferocity as those who claimed the pertussis vaccine worked. Stewart believed AIDS was caused by the gay lifestyle, not HIV. In 1995, more than ten years after the discovery and confirmation of HIV as the cause of AIDS, Stewart wrote, "AIDS and AIDS-related complexes develop, with and without HIV, because [foreign proteins] in spermatozoa enter the rectum and bloodstream . . . and elicit antibodies which are toxic to [white blood cells]." Stewart didn't believe HIV caused immune deficiency; he believed sperm in the rectum did it. He also believed yeast infections in homosexual men did the same thing. Stewart blamed the victim, not the virus, writing: "Every time an avowedly homosexual or bisexual rock or film star dies of the disease he is elevated to martyr and hero. Yet it is an unpalatable and unpopular fact, seldom articulated, that those who die of AIDS, like the smokers who die of lung cancer and heart disease, facilitated their own death." Stewart concluded, "We should take the sentimentality out of AIDS and recognize that the disease is, with few exceptional cases, directly caused by the behavior of the victim. If we do that, it would be better for all concerned."

In 1995, Stewart also argued against giving azidothymidine, an anti-viral medication, to pregnant women with AIDS. His plea came at a time when AZT had already been shown to prevent transmission of HIV from pregnant women to their unborn babies. His unconscionable rants against AIDS victims, his ill-founded notions about the cause of the disease, and his strident campaign against a valuable drug made Gordon Stewart a target of ridicule.

These were the men Lea Thompson chose to educate the American public about pertussis vaccine.

During *Vaccine Roulette*, Lea Thompson interviewed a personal-injury lawyer named Allen McDowell. McDowell claimed that he had uncovered a conspiracy among doctors to hide the truth about pertussis vaccine. "[In] some institutions that I've seen in this state [Illinois]," said McDowell, "certain administrators . . . have indicated that they have children there as a result of the DPT. Brain-damaged children." And he believed that it wasn't only doctors who covered up the fact that the damage had been caused by the vaccine; vaccine makers were in on it, too. Thompson said that attorneys had accused "the vaccine manufacturers of destroying vaccine records before they [could] be subpoenaed for a DPT lawsuit." Then Thompson served the home run pitch—the kind of question personal-injury lawyers can only dream about. "Do you think that some children have been damaged by the DPT shot and their parents don't even know it?" she asked, arguing that the few patients she had described were only the tip of an iceberg—an iceberg of children whose parents could ask Allen McDowell to represent them. "Absolutely," enthused McDowell. "I don't think the parents would be aware of [the alleged vaccine harm] and normally the pediatrician, or whoever, the GP, wouldn't tell them." Later, Thompson concluded, "What about the children who have already been damaged—who's helping them? Unless they sue—and many families don't have the money or don't want to do that—nobody is helping them to pay the enormous costs that a brain-damaged child brings upon the family." "A child like [Scottie] deserves to stay out of an institution," said Marge Grant, seemingly in response to McDowell. "And, unless there's compensation, you simply cannot do it." If personal-injury lawyers had wanted to make an infomercial on alleged harm caused by pertussis vaccine, they couldn't have done much better than *Vaccine Roulette*.

In 2007, John Stossel interviewed Allen McDowell on an *ABC News* program titled "Scared Stiff: Worry in America." During the program, Stossel showed a clip of news reporter Robin Roberts and Allen McDowell during the pertussis vaccine scare. "Sudden infant death," said Roberts. "You put your baby into the crib, you wake up the next morning, and your baby is dead." "It's extremely dangerous," said McDowell. "They had the ability to make a safer vaccine." Twenty-five years had passed since Lea Thomspon's program.

During that time, epidemiological studies had clearly shown that pertussis vaccine didn't cause brain damage or Sudden Infant Death Syndrome, and advances in neurology and genetics had better defined the real causes of the problems. But Allen McDowell was unbowed. "The vaccine just wasn't as safe as it could have been?" asked Stossel. "There's no dispute about that," replied McDowell. "They were making so much money off the old vaccine that they didn't really have any incentive to improve it." Stossel, in voice-over, said, "McDowell made money, too. The lawyer who now works out of his house won lots of lawsuits—for how much money?" "I made a good chunk of money," said McDowell. "One hundred million?" asked Stossel. "I really can't say," McDowell demurred. "It's under protective order."

During his interview with Stossel, McDowell never mentioned his law partner, Anthony Colantoni. Between July 1990 and October 1991, Colantoni received $1.53 million in compensation checks from the Vaccine Injury Compensation Program for families of children presumably damaged by pertussis vaccine. But Colantoni gave only $124,000 to his clients, putting the remaining $1.4 million into the account of McDowell & Colantoni. His method of concealing what he'd done was simple. "When the victims or family members of the estates [of plaintiffs] would call," said U.S. Attorney James B. Burns, "he would deny he had received the checks." In March 1993, by consent of the Illinois Supreme Court, Anthony Colantoni was disbarred. McDowell was never implicated in his partner's scheme.

Vaccine Roulette was arguably one of the most powerful programs ever to air on American television: thousands of parents stopped giving pertussis vaccine to their children; personal-injury lawyers pummeled pharmaceutical companies, causing many to stop making vaccines; and Congress passed a law to protect vaccine makers, while at the same time compensating those who were allegedly harmed by vaccines.

During the next fifteen years the tide turned. David Miller's study lay in ruins at the hands of a British judge. And study after study

showed that children immunized with DTP weren't at greater risk
of brain damage. As a consequence, public health agencies and medical societies throughout the world no longer considered pertussis
vaccine to be a rare cause of permanent harm. Even the Vaccine Injury Compensation Program, a system designed to compensate those
who felt wronged by vaccines, had removed epilepsy as a possible
consequence of pertussis vaccine.

Despite this overwhelming evidence, and despite all of the harm
that had been done by the false notion that pertussis vaccine was
maiming America's children, Lea Thompson was without remorse.
In 1997, during a celebration in her honor held by the National Vaccine Information Center, the group once called Dissatisfied Parents
Together, Thompson remembered *Vaccine Roulette*: "The reason it
was important to me is not because it was great research, although
we did a pretty good job, or that [it] was a beautifully produced
piece of work. *DPT* [*Vaccine Roulette*] was important to me personally because it spawned a movement." A movement that almost
eliminated vaccines for American children, a movement that continues to cause many parents to reject vaccines in favor of the diseases they prevent, and a movement that was based on a notion that
has been shown again and again to be incorrect.

In retrospect, it isn't surprising that an anti-vaccine movement sprang
up in the United States in the 1980s. The surprise is that it didn't
happen sooner.

In the early 1940s, a yellow fever vaccine was routinely given to
American soldiers. Everyone got it. To make sure the vaccine virus
was stable across a broad range of temperatures, manufacturers added
human serum, a decision that proved disastrous. (Serum is blood
without red blood cells or clotting factors.) Unbeknownst to the manufacturer, some of the blood donors had hepatitis. At the time, scientists didn't know about the different types of hepatitis viruses or how
they were spread. In March 1942, the Surgeon General's Office noted
a striking increase in the number of recruits with hepatitis; more than
three hundred thousand soldiers were infected with what we now
know as hepatitis B virus; sixty-two died from the disease.

In the early 1950s, Jonas Salk made a vaccine to prevent polio. Horrified that children could be fine one minute and wheelchair-bound the next, Americans gave their money to the March of Dimes, which gave it to Jonas Salk. Salk reasoned that killed poliovirus would induce a protective immune response without causing disease. Working with mice and then monkeys, Salk figured out how to make a vaccine by growing poliovirus in laboratory cells, purifying it away from the cells, and killing it with formaldehyde. In 1954, the March of Dimes tested Salk's vaccine in a trial of almost two million children. (It was then and remains today the largest vaccine trial ever performed.) When the results of the study were announced, church bells rang across the country, factories observed moments of silence, synagogues held special prayer meetings, and parents, teachers, and students wept. "It was as if a war had ended," one observer recalled. The euphoria didn't last long. Two weeks later, public health officials recalled every lot of polio vaccine.

When the March of Dimes had tested Salk's vaccine, it relied on two veteran vaccine makers: Eli Lilly and Parke-Davis. But when the vaccine was licensed for sale, three other companies joined in: Wyeth, Pitman-Moore, and Cutter Laboratories. It soon became clear that Cutter, a small pharmaceutical company in Berkeley, California, had made it badly, failing to fully inactivate the virus. As a consequence, one hundred and twenty thousand children were inadvertently injected with a vaccine that contained live, potentially deadly poliovirus: seventy thousand suffered mild polio, two hundred were severely and permanently paralyzed, and ten died. It was one of the worst biological disasters in American history. Cutter Laboratories never made another dose of polio vaccine again.

Perhaps the worst vaccine disaster in history didn't occur in America; it occurred in Germany. In 1921, two French researchers, Albert Calmette, a physician, and Camille Guérin, a veterinarian, reasoned that a bacterium (*Mycobacterium bovis*) that caused tuberculosis in cows could protect people against human tuberculosis. They developed a vaccine later called BCG (Bacillus of Calmette and Guérin), a modified form of which is still used today. In 1929, however, 250 ten-day-old children in Lubeck were given a BCG vaccine that wasn't made of BCG. It was made of pure, highly lethal human tuberculosis bacteria. Seventy-two babies died from the mistake.

The yellow fever, Cutter, and BCG vaccine disasters didn't spur significant anti-vaccine activity. (However, the Cutter incident led to the creation of a vaccine regulatory system that prevented these kinds of tragedies from happening again.) In America, people still trusted vaccines; and they trusted those who made and recommended them. It would take fear of pertussis vaccine to turn the tide; ironic, given that the pertussis vaccine tragedy was imagined.

CHAPTER 5

Make the Angels Weep

If they can get you asking the wrong questions, they don't have to worry about the answers.
—THOMAS PYNCHON, *GRAVITY'S RAINBOW*

By the late 1980s, Barbara Loe Fisher was riding high. She had written *A Shot in the Dark: Why the P in the DPT Vaccination May Be Hazardous to Your Child's Health*, praised by the *San Francisco Chronicle* as "cautious, credible, horrifying, and outrageous all at once." She had spurred an enormous effort by academic researchers, pharmaceutical companies, and public health officials to make a purer pertussis vaccine. She had helped craft legislation that included a monitoring system for licensed vaccines—a system that would, ten years later, detect a rare but serious side effect. And although the reason for Fisher's activism—her belief that pertussis vaccine had caused her son's learning disabilities—wasn't supported by the science, she had been a catalyst to changes that clearly benefited children. She was, in short, America's premier vaccine safety activist. The media believed her, politicians relied on her, and parents turned to her. Barbara Loe Fisher was poised to do a tremendous amount of good.

Unfortunately, during the next three decades, the opportunity was squandered—an opportunity that had been so hard won.

When Barbara Loe Fisher burst onto the scene, several vaccines had serious side effects, every year causing allergic reactions, paralysis, or death. Public health officials and doctors didn't hide these problems. But they didn't do anything to correct them, either. And most parents had no idea they existed.

Beginning in the early 1960s, American children were given a polio vaccine that was ingested, not injected. Albert Sabin, a well-respected virologist and Jonas Salk's fiercest rival, invented it. Sabin's approach was dramatically different from Salk's; instead of killing poliovirus with a chemical, as Salk had done, he weakened it. Sabin reasoned that by taking poliovirus and growing it over and over again in nonhuman cells, the virus would become less and less capable of reproducing itself in humans. And he was right. Sabin's vaccine, dropped onto sugar cubes and given to millions of American children, worked. By 1979, polio, a disease that had caused hundreds of thousands of children to suffer and die, was eliminated from the United States. By 1991, it was eliminated from the Western Hemisphere—a remarkable accomplishment.

But there was one problem.

When Albert Sabin weakened poliovirus in his laboratory, he found that it could no longer grow in the brains and spinal cords of experimental monkeys; so he reasoned that the vaccine virus wouldn't grow in children, either. But Sabin hadn't anticipated a rare occurrence: polio caused by his polio vaccine. Although this problem was extremely rare—occurring in 1 of 2.5 million doses—it was real. Every year for the next twenty years, six to eight children in the United States got polio from the oral polio vaccine. And some of these children died from the disease. The problem caused by Sabin's vaccine was avoidable; several countries never used it, relying on Salk's to successfully eliminate the disease.

When Barbara Loe Fisher became a vaccine safety activist, she could have taken on Albert Sabin's polio vaccine. Pharmaceutical companies had little incentive to make an inactivated polio vaccine—one that didn't occasionally cause paralysis—and public health officials were unwilling to spend more money on a polio vaccine in the absence of public demand. (Because it required a syringe and needle, as well as a medical professional to give the shot, Salk's vaccine was much more expensive than Sabin's, which could just be squirted into the mouth.) It was a perfect situation for a consumer advocate. Years later, one advocate would force the government to acknowledge the rare but invariant paralysis that came with Sabin's polio vaccine and to change public policy. It could have been Barbara Loe Fisher. But it wasn't.

Sabin's polio vaccine wasn't the only problem.

Ten years before Jonas Salk made his polio vaccine, his former mentor, Thomas Francis, made an influenza vaccine. Francis took influenza virus, injected it into eggs, grew the virus, purified it, and inactivated it with formaldehyde. (Salk's idea of inactivating poliovirus with formaldehyde came from working in Francis's laboratory.) The influenza vaccine is made the same way today. Unfortunately, some people can't get it because they're severely allergic to eggs. (In the United States about a million are.) Reactions can be frightening, and can include hives, low blood pressure, difficulty breathing, and shock—all of which could be avoided; companies could grow influenza viruses in mammalian cells rather than avian ones. Although this procedure wouldn't be easy, it's doable. But, absent a public outcry, pharmaceutical companies have had little incentive to make the change and public health agencies haven't insisted they do it. Again, it's a perfect situation for an advocate.

Egg proteins aren't the only vaccine component that causes severe allergic reactions, or even the most common cause of them; gelatin is. Made by extracting collagen from the bones and hides of pigs, gelatin is used as a stabilizing agent, allowing small quantities of live viral vaccines to be evenly distributed throughout a vial. (For decades, the MMR vaccine contained gelatin as a stabilizer; today only the chickenpox and nasal-spray influenza vaccines have it.) Most people don't have a problem with gelatin, but some develop severe allergic reactions. Also, certain religious groups are hesitant to receive a vaccine made with pig products. Again, for those wanting to make vaccines safer, gelatin would be a great place to start. Other stabilizing agents are available.

Barbara Loe Fisher
founded America's
modern-day anti-vaccine
movement. (Courtesy of
Dayna Smith.)

All of these problems were worthy targets for a consumer advocate. But Barbara Loe Fisher chose to take vaccine safety activism in a dramatically different direction.

Beginning in the early 1980s, and for the next three decades, every time a new vaccine was recommended, the media sought out Barbara Loe Fisher for her opinion. The first vaccine licensed under Fisher's watch prevented Hib.

Older American pediatricians have witnessed the horror of Hib disease. Before the vaccine, Hib was the most common cause of meningitis, often leaving children deaf, blind, or severely mentally disabled. And it was a common cause of bloodstream infections (sepsis) and pneumonia. But one disease caused by Hib is even more frightening—a disease that most parents have never heard of and most young doctors have never seen: epiglottitis.

The epiglottis is a thumb-like wedge of tissue at the back of the throat; when people swallow, it flops down over the windpipe, preventing food and water from entering the lungs. Hib is unique among bacteria in its capacity to infect the epiglottis. Once infected, the epiglottis can block the windpipe—no different, in a sense, than being smothered by a pillow. Before 1990, every big city hospital had an "epiglottitis team," designed to usher children quickly and quietly into the operating room for a lifesaving tracheostomy (a surgical hole in the windpipe). The *quietly* part was particularly important. Once agitated, children with epiglottitis were much more likely to suffocate on the spot. No disease was more nerve-wracking.

In 1987, the Food and Drug Administration licensed the first Hib vaccine. To doctors across the country, the vaccine was a godsend. At last, they could prevent the severe, permanently disabling, and fatal cases of Hib that occurred every year. Barbara Loe Fisher didn't share their enthusiasm. On *World News Tonight with Peter Jennings*, she cautioned, "We have to do more independent vaccine risk studies to find out whether or not vaccinations are causing chronic diseases, like diabetes." Fisher's response was predictable. Years before, in her book *A Shot in the Dark*, she had written, "With the increasing number of vaccinations American babies have been required to use has come increasing numbers of reports of chronic immune and neurological disorders being suffered by older children and young adults including asthma, chronic ear infections, autism, learning disabilities, attention deficit disorder, diabetes, rheumatoid arthritis, multiple sclerosis, chronic fatigue syndrome, lupus, and cancer. The unanswered question is whether multiple vaccinations, which are suppressing many diseases, especially in childhood, are playing a leading role in the rise of chronic illnesses later in life." By simply replacing infectious diseases with chronic diseases, vaccines, according to Fisher, were responsible for many of mankind's ills. For every new vaccine, Fisher would find at least one doctor to support her view. In the case of the Hib vaccine, it was Bart Classen.

Classen was the president and chief executive officer of Classen Immunotherapies, a company that held patents on alternative vaccine schedules and alternative methods to identify vaccine side effects. To establish that Hib vaccine caused diabetes, Classen showed that children in Finland who had received three doses of vaccine in infancy were more likely to have diabetes than those who had received only one dose in the second year of life. Appearing with Barbara Loe Fisher on *World News Tonight*, he said, "In fact, the scheduling has a major impact on the development of diabetes.

We're saying that when you look at the big picture, looking at five, ten years down the road, that this isn't the ideal schedule."

This was big news. The Hib vaccine, thought to be one of medicine's greatest lifesavers, was, according to Classen, actually causing children to suffer a lifelong, debilitating, often fatal disease. Other researchers rushed to confirm Classen's findings. One group examined the risk of diabetes in twenty-one thousand American children who had received the Hib vaccine and compared it with twenty-one thousand children who hadn't. The risk of diabetes was the same in both groups. Other investigators examined two hundred and fifty American children with diabetes to see if they were more likely to have been vaccinated than other children. Again, Hib vaccine didn't increase the risk of diabetes. Indeed, no vaccine has ever been shown to increase the risk of diabetes. The inability of researchers to reproduce Classen's findings caused them to take a closer look at his original study. They found severe flaws in his analytical methods, further substantiated when, ten years after vaccination, Finnish children who had received Hib vaccine in infancy weren't more likely to have diabetes—exactly the opposite of what Classen had said on *World News Tonight*.

Barbara Loe Fisher's appearance on national television with Bart Classen wasn't her only public comment on the Hib vaccine. One ironic twist remained.

Heather Whitestone was born in 1973 in the tiny town of Dothan, Alabama. After high school Whitestone attended Jacksonville State University and started competing in beauty pageants. On September 17, 1994, at the age of twenty-one, she was crowned Miss America— the first with a severe disability. "I lost my hearing when I was eighteen months old," she said. After Whitestone accepted her crown, her mother told a local news reporter what she thought had caused her daughter's deafness: DTP vaccine. Whitestone's prominence, combined with her mother's revelation, was chilling for any parent deciding whether to vaccinate their child. Barbara Loe Fisher quickly weighed in: "It's so often that the parent who is with the child and witnesses the high fever, witnesses the convulsions, the shock or whatever; it comes down to whether the doctor agrees it was due to the

Heather Whitestone, the first Miss America with a disability, was a controversial figure in the anti-vaccine movement. (Courtesy of Donna Connor/Sygma/Corbis.)

vaccination just given. [The medical community] continues to try to sweep these children under the rug."

Heather's mother had omitted one critical part of the story. Ted Williams, the Dothan, Alabama, pediatrician who had taken care of Heather, had followed her rise to fame with pride. But when Heather's mother claimed that DTP had caused her daughter's illness, Williams stepped forward. He knew that Heather wasn't deaf because of DTP; she was deaf because of a near-fatal case of Hib meningitis. Fisher responded to Williams's revelations angrily, seeing conspiracy. "Within hours after the Miss America pageant, a horrified medical establishment moved quickly to publicly dispute any connection between Heather's deafness and the DPT vaccine and instead blame her deafness on a bacterial infection," said Fisher. "The American medical establishment went to extraordinary lengths to publicly challenge Heather and her mother in order to avoid having to acknowledge DPT vaccine risks."

American children have used Hib vaccine for more than twenty years. During that time the number with Hib meningitis, bloodstream infections, pneumonia, and epiglottitis has decreased from twenty thousand every year to fewer than fifty. Sadly, some parents watching *World News Tonight*, frightened by Barbara Loe Fisher's and Bart Classen's warning that Hib vaccine could cause diabetes, might have chosen not to vaccinate their children: a choice that would have put them at unnecessary risk of a highly disabling, often fatal infection.

The next vaccine recommended for infants protected against hepatitis B virus. It was immediately controversial. Although most parents had probably never heard of Hib, they had all heard of and feared meningitis. So the Hib vaccine was an easy sell. But while most parents know what hepatitis is, they had never considered it a disease of children. Actually, it's more common than people realize.

Before the vaccine, hepatitis B virus infected about two hundred thousand people in the United States every year, mostly teenagers and young adults. The infection isn't trivial, causing fever, vomiting, nausea, food intolerance, abdominal pain, headache, muscle pain, joint pain, rash, and dark urine followed a few days later by jaundice (yellowing of the skin and eyes) and an enlargement and tenderness of the liver. Some patients become disoriented, extremely sleepy, semi-conscious, or comatose: all symptoms of severe liver damage. Four of ten people in the United States with symptomatic hepatitis B infection will be hospitalized and five thousand killed by the virus every year. For those who survive, the virus may cause a long-lived infection, resulting in permanent scarring of the liver (cirrhosis) or liver cancer. Worse: people chronically infected with hepatitis B virus often don't know it—problematic, given that many are highly contagious to others. For this reason, hepatitis B infections are known as the "silent epidemic." Prior to the development of hepatitis B vaccine, about a million people in the United States were chronically infected.

In 1981, the year before Barbara Loe Fisher watched *Vaccine Roulette*, the first hepatitis B vaccine was licensed. It wasn't recommended for routine use in children. Public health officials reasoned that the best way to eliminate the disease was to recommend the vaccine for those most likely to become infected—specifically, people who have sex with an infected person, especially men who have sex with men; healthcare providers unknowingly exposed to contagious patients; intravenous drug users; prisoners; and people who get tattoos from places that don't adequately sterilize equipment between customers. Between 1981 and 1991, the vaccine was recommended only for those at highest risk. Unfortunately, the strategy failed, miserably; and the incidence of hepatitis B virus infections remained unchanged. So, government officials embarked on the second stage of their plan, recommending three doses of the hepatitis B vaccine for all babies, the first to be given soon after birth.

The new hepatitis B vaccine policy was a public relations nightmare. Because the vaccine had initially been recommended for prisoners, intravenous drug users, and men who had sex with men, it was viewed as "dirty": a vaccine that had no place in the infant vaccine schedule. Fisher leveraged this perception to promote her premise that vaccines caused chronic diseases. The doctor who supported her this time was Bonnie Dunbar.

On January 22, 1999, ABC's *20/20* aired a program that deeply scared the American public. (Fisher, who provided information to the show's producer, can be seen briefly pointing to a computer screen.) Sylvia Chase was the correspondent. The program began with a teaser: "Next, an important medical controversy. Serious new questions about a vaccine most school children are forced to get: one given to millions of babies every year." "We just thought it was like all the immunization shots," said one mother. "We were doing it to protect our child." "It's the hepatitis B vaccine," warned the announcer. "These parents thought it would protect their child. No one told them that there might be risks." "Within three weeks of the third shot, I lost my vision," said one woman. The announcer then revealed the show's premise: "Is it smart preventive medicine, or an unnecessary risk? Sylvia Chase asks: When it comes to hepatitis B, who's calling the shots?"

The *20/20* program told the stories of several healthcare providers who had developed multiple sclerosis after getting the hepatitis B vaccine.

CHASE: These medical workers say they were healthy. Then they were vaccinated.

WORKER #1: Within two months I was very ill. I was ill in bed.

WORKER #2: I can't even feed myself.

DUNBAR: These people were completely healthy.

CHASE: Dr. Bonnie Dunbar, a cellular biologist at Baylor College of Medicine, believes that in certain people a genetic component sets off an explosive chain of events.

DUNBAR: The only thing that happened is they took this vaccine and within a month most of these people had completely debilitating life-style changes.

The implication was clear; hepatitis B vaccine caused multiple sclerosis. Chase then described the same problem that Lea Thompson had described for DTP:

CHASE: Three-day-old Ben Converse's seizures began less than twenty-four hours after his first shot. Now Ben is developmentally disabled. Thirty-three hours after his vaccination, thirteen-day-old Nicky Sexton's heart stopped. The coroner said it was Sudden Infant Death Syndrome, or SIDS. Lyla Belkin's death was also attributed to SIDS. She had received her first shot at six days old: the second one, a month later.

MR. BELKIN: How is a baby possibly going to get hepatitis B? It's ridiculous to give this vaccine to a child. I wish we'd known that before receiving this vaccine.

CHASE: You did what the doctor told you to do.

MRS. BELKIN: I did what the doctor told me to do. Yes, of course.

CHASE: And they didn't say that there were any cases of deaths or serious reactions?

MRS. BELKIN: No, nothing.

CHASE: On September 16, 1998, Mrs. Belkin nursed Lyla at 5:30 a.m., not long after, she found her pale and cold.

MRS. BELKIN: (fighting back tears) She died early in the morning about sixteen hours after vaccination.

The scene shifted back to the show's hosts, Hugh Downs and Barbara Walters. Walters looked into the camera, incredulous. "What a choice for parents to have to make," she declared. "There is so much conflicting information!" Unfortunately, the problem with the program wasn't that there was so much conflicting information; it was that there was so much *wrong* information, such as the false notion that babies aren't at risk of getting infected. "How's a baby possibly going to catch hepatitis B virus?" Michael Belkin had asked. Fisher echoed Belkin's disbelief. "There are only four hundred cases a year of hepatitis B in children under fourteen [years of age]," she said. "It's not a disease your average, healthy child is likely to contract." Although it is true that most disease and death caused by hepatitis B virus occur in adults, every year before the hepatitis B vaccine about sixteen thousand children less than ten years of age were infected by nonsexual, person-to-person contact. (Hepatitis B virus can spread fairly casually, such as by sharing toothbrushes.) Worst of all, infants infected with hepatitis B virus are at the highest risk of long-term problems; many develop cirrhosis or liver cancer later in life. Although childhood infections accounted for fewer than 10 percent of all infections in the United States, they accounted for 20 percent of all cases of chronic liver disease. This is the reason public health officials had recommended the vaccine for newborns.

Another misleading message from the *20/20* broadcast was that hepatitis B vaccine caused SIDS. The most compelling story was that of Lyla Belkin, who died of SIDS following her second dose. Fisher asked Michael Belkin to head the Hepatitis B Vaccine Project at her National Vaccine Information Center. Soon Belkin, a Wall Street financial advisor, was everywhere. In February 1999, at a meeting of a federal vaccine advisory group at the CDC, he said, "I hold each one of you who participated in the promulgation or perpetuation of that mandated newborn vaccination policy personally responsible for the death of my daughter." Several months later, Belkin told a congressional committee, "Almost every newborn U.S. baby is now greeted on its entry into the world by a vaccine injection against a sexually transmitted disease for which the baby is not at risk—because they [health officials] couldn't get the junkies, prostitutes, homosexuals, and promiscuous heterosexuals to take the vaccine. Parents need to understand that the system providing the vaccines injected into their children's veins is corrupt and scientifically flawed."

Despite Belkin's certainty that hepatitis B vaccine had caused his daughter's SIDS, study after study failed to support him. Indeed, during the 1990s, when the vaccine was given to more and more babies, fewer and fewer died of SIDS. This wasn't because of the vaccine; it was because of an aggressive program called "Back to Sleep." Investigators had found that children who died of SIDS were more likely to have been lying on their faces than on their backs at the time of death. Encouraging parents to lay their children on their backs dramatically reduced the incidence of SIDS.

Sylvia Chase's *20/20* segment also trumpeted the notion that hepatitis B vaccine caused multiple sclerosis—an accusation that didn't hold up. Two years after the program, investigators published two large studies that assessed whether the vaccine caused multiple sclerosis (it didn't) or worsened symptoms in people who already had the disease (again, it didn't).

At the time of the *20/20* broadcast, the hepatitis B vaccine had been given to fifty million infants and children and seventy million teenagers and adults in the United States. Since 1991, when it was first recommended for all infants, the vaccine has virtually eliminated the disease in children. Barbara Loe Fisher used her considerable platform in the 1990s to warn American parents that the hepatitis B vaccine was unnecessary, that it caused SIDS, and that it caused multiple sclerosis—positions that could have misled parents into avoiding a vaccine that has prevented a great deal of suffering and death.

Fisher's assault on newly recommended vaccines didn't end with hepatitis B. Next up was a vaccine to prevent pneumococcus, the most common cause of pneumonia and, like Hib, an important cause of meningitis and bloodstream infections.

In 1998, researchers in Northern California performed a land-mark study. They tested the pneumococcal vaccine in thirty-eight thousand infants: half received it and half didn't. The results were dramatic. Pneumococcus caused bloodstream infections in seven-teen children who didn't get the vaccine and in not a single child who did: seventeen versus zero. Excited, researchers made their findings public.

On the evening of September 25, 1998, John McKenzie, a cor-respondent for *World News Tonight with Peter Jennings*, told the story of the pneumococcal vaccine. The program included Barbara Loe Fisher, Bart Classen, and Neal Halsey, a vaccine expert from the AAP.

JENNINGS: In California today, researchers have announced a new vaccine which they say will protect children from several diseases. They are very encouraged about the clinical trials. But the news raises a question—do children need another vaccine? Here's ABC's John McKenzie.

MCKENZIE: Most children in this country are injected with at least ten different vaccines. Now, some doctors seem eager to add yet another—this new pneumococcal vaccine.

NEAL HALSEY: It's one of the most important, if not the most im-portant, vaccines that has been developed in the last ten years.

MCKENZIE: The vaccine appears to prevent the most dangerous form of pneumococcal disease called bacterial meningitis. It also might help prevent pneumonia and middle ear infection. While few people doubt that this new vaccine is effective, a controversy is erupting about whether every child in the country actually needs it, whether we really know enough to say the benefits outweigh the risks. Young children are the most vulnerable to serious pneu-mococcal disease, with ten thousand cases reported each year. But relatively few, only about two hundred, are actually deadly. The vast majority are treated effectively with antibiotics.

FISHER: If we're going to require a vaccine, it should be for a dis-ease that's highly contagious, is extremely deadly, and is in epi-demic form. This disease does not qualify.

MCKENZIE: The reason that some are urging caution is that nobody knows the long-term side effects of any vaccine. And the few cases where people have started to look, the results have been disturbing.

CLASSEN: Published studies showed that when one follows these kids, and looks for diseases like diabetes and asthma . . . the immunized kids seem to be at an increased risk.

FISHER: The vaccine should definitely be made available for children who are in poor health, having a compromised immune system, but it should not be required for all children.

MCKENZIE: While vaccines have saved the lives of millions, some researchers warn that before we add another, we have a better understanding of the risks involved.

At the time that the *World News Tonight* piece aired, parents didn't have a choice about whether or not to get the pneumococcal vaccine. That's because it hadn't been licensed yet. But for parents who would soon face that choice, *World News Tonight* had done much to mislead them.

As she had done with the Hib vaccine, Fisher allied herself with Bart Classen, who, without any supportive evidence, claimed that the pneumococcal vaccine caused diabetes. Also, Fisher had claimed that diseases caused by pneumococcus were not particularly common, affecting only those in poor health. This wasn't true. Before the vaccine, every year pneumococcus caused four million ear infections, one hundred and twenty thousand cases of pneumonia requiring hospitalization, thirty thousand bloodstream infections, and twenty-five hundred cases of meningitis. Most of these diseases occurred in previously healthy children. And, although the bacterium didn't kill as many children as measles or polio, John McKenzie's comment that it killed "relatively few, only about two hundred" was rather callous to those parents whose children had suffered and died from pneumococcus.

Within a year of the *World News Tonight* program, federal health officials asked Barbara Loe Fisher to serve on one of the most powerful vaccine advisory committees in the United States—a committee that advised the FDA. No vaccines have been licensed without approval from this committee. Fisher was asked to serve because health officials believed that once she saw how carefully vaccines were tested—once she got a look behind the curtain—she would feel more confident about the safety of vaccines. It didn't happen.

The first vaccine on which Fisher was asked to vote was the pneumococcal vaccine. In 1999, after reviewing all of the safety and efficacy data, committee members cast their vote: 11 to 1 in favor of licensing the vaccine. The lone dissenting vote was that of Barbara Loe Fisher. Because the FDA didn't require unanimous approval for licensure, in January 2000 it licensed the vaccine. Robert Daum, a professor of pediatrics at the University of Chicago Children's Hospital, said, "I think this is a giant step for the health of children." Fisher disagreed. "There's not enough evidence about the safety of this vaccine," she said. "What we basically have here is a post-marketing experiment."

By 2009, almost a hundred million doses of the pneumococcal vaccine had been given to American children. As a consequence, the incidence of pneumococcal disease has decreased dramatically.[37] Far fewer children now get meningitis, pneumonia, and bloodstream infections caused by pneumococcus. And, although we'll never know their names, hundreds of children are still alive because they got the pneumococcal vaccine. That's not all. Because the vaccine has reduced the number of children who carry pneumococcus in the nose and throat, older people, such as their grandparents, have also benefited. Finally, the chronic diseases predicted by Barbara Loe Fisher and Bart Classen in front of millions and millions of television viewers never materialized.

In 1998, another vaccine was licensed and recommended for young children. After the vaccine had been used for about a year, something happened that should have put Barbara Loe Fisher's concerns about the government's interest in vaccine safety to rest.

On August 31, 1998, the FDA licensed a vaccine that protected against a common intestinal virus called rotavirus. Rotaviruses, which cause fever, vomiting, and diarrhea in infants and young children, are responsible for nine hundred thousand office visits, seventy thousand hospitalizations, and sixty deaths every year in the United States, mostly from dehydration. In the developing world, rotavirus is a more prodigious killer, causing the deaths of two thousand children every day. Because the disease is common and occasionally fatal, the CDC recommended the vaccine for all infants.

In July 1999—ten months after licensure—the CDC discovered something it hadn't anticipated. Reports of fifteen children who had received rotavirus vaccine appeared in the Vaccine Adverse Events Reporting System (VAERS). These children shared several features: all had developed an uncommon form of intestinal blockage called intussusception (which occurs when one segment of the small intestine telescopes into another and gets stuck), all had recently received the rotavirus vaccine, and most were about two months old (intussusception is unusual in two-month-old children). Intussusception is a medical emergency. Children with the disorder can develop severe bleeding from the intestine or a bloodstream infection; both can be fatal.

The CDC knew that it had a problem. So, on July 16, 1999, it temporarily suspended the use of the new rotavirus vaccine until investigators could figure out what was going on. Finding an answer wasn't going to be easy. Every year before the rotavirus vaccine had become available, one in two thousand infants in the United States developed intussusception, most between five and nine months of age. Jeff Koplan, head of the CDC, pulled people off other projects, spent millions of dollars, and made it clear that an explanation for the intussusception found in two-month-olds must come quickly. By October, only months after the possible association between rotavirus vaccine and intussusception was reported to VAERS, Koplan and his CDC team figured it out. Children who received the new rotavirus vaccine were twenty-five times more likely to get intussusception than those who hadn't. Although the rotavirus vaccine clearly caused intussusception, the risk of getting the disease was quite low: one case per ten thousand vaccine recipients. That same month, the CDC withdrew its recommendation of the vaccine—and the company that produced it took it off the market. Seven years would pass before a safer rotavirus vaccine was made.

The serious side effect caused by the first rotavirus vaccine offered several insights into how the CDC monitors vaccines. Reports to VAERS quickly signaled a problem; CDC officials acted immediately, withholding the vaccine until they determined whether the problem was real and, once they found it, withdrew their recommendation. This is what happens when a vaccine actually causes a problem. Barbara Loe Fisher should have been reassured by all of this. But she wasn't.

Another vaccine licensed on Fisher's watch, and one that incurred her greatest wrath, prevented human papillomavirus (HPV), a known cause of cancer of the cervix. Cervical cancer isn't rare, causing ten thousand American women to suffer and four thousand to die every year. The bad news is that about thirty different strains of HPV cause cancer. The good news is that two of the strains contained in the vaccine prevent 70 percent of cases. And HPV isn't one of many viruses that cause cervical cancer; it's the only virus that causes it. So, the HPV vaccine was a lifesaving breakthrough.

Although HPV vaccine was new, the strategy used to make it wasn't. It was made with the same technology used to make the hepatitis B vaccine twenty years earlier. By 2006—when the CDC recommended it for teenage and adult women—HPV vaccine had been tested for seven years in more than thirty thousand women. Other than pain and tenderness at the site of injection, and occasional episodes of fainting, the vaccine didn't appear to have any serious side effects. But Barbara Loe Fisher was determined to defeat the HPV vaccine, again claiming that it wasn't necessary and that pharmaceutical companies had misrepresented data. "Merck and the FDA have not been completely honest with the people about the prelicensure clinical trials," said Fisher. "Merck's pre- and post-licensure marketing strategy has positioned mass use of this vaccine by preteens as a morality play in order to avoid talking about the flawed science they used to get licensure."

Two months later, Fisher turned up the volume: "In what is perhaps the most brilliant public relations and marketing strategy ever employed by a pharmaceutical company promoting the universal use of a vaccine most Americans do not need to prevent cervical cancer, Merck and Co. is in the process of pulling off one of the biggest money-making schemes in the history of medicine. The Big Pharma giant that brought us death by Vioxx has convinced the FDA, CDC, public-school officials, and gynecology professors as well as the entire European Union that every man and woman in the world must purchase and be injected with the HPV vaccine in order to survive." Later, on her blog, she called it "the slut shot" and "the 'cheaters' vaccine." (Because HPV infects 70 percent of women within five years of their first sexual encounter, one could infer that Barbara Loe Fisher doesn't have a very high opinion of American women.)

As she had with the Hib, pneumococcal, and hepatitis B vaccines, Fisher claimed that HPV vaccine caused chronic, serious disabilities. She reported that young girls had been paralyzed and were dying from the vaccine: dying from blood clots that caused strokes and heart attacks. On CBS's *Sunday Morning*, Fisher told Charles Osgood and millions of television viewers, "This is an intervention that carries the risk of injury or death." CDC officials responded to the public's growing concern, studying more than ten thousand reports to VAERS to determine if there was a pattern (as had been seen with the ill-fated first rotavirus vaccine) that suggested HPV vaccine might be causing a problem. They examined reports of paralysis and found "there has been no indication that [HPV vaccine] increases the rate of Guillain-Barré Syndrome [GBS, a rare cause of paralysis] in girls and women above the rate expected in the general population." They also examined the medical records of twenty-seven people who had died soon after getting the vaccine, finding nothing to suggest that the vaccine had been the cause. People who had recently received HPV vaccine had died from complications of diabetes or heart failure or viral illnesses or bacterial meningitis or drug overdose or blood clots caused by birth-control pills or seizures in those already known to have epilepsy. Reports of death following HPV vaccine were consistent with deaths that were occurring in the general population before the vaccine was used. Fisher didn't believe it. "The 'coincidence' defense mounted by doctors and drug-company officials every time a vaccination is followed by injury and death is as old as it is unscientific," she blogged. "It's amazing that they have been able to get away with it for so long."

Perhaps Fisher's most disingenuous comment about HPV vaccine was that it might cause cancer: "And how many [girls] will go on to develop fertility problems, cancer, or damage to their genes, all of which Merck admits in its product insert it hasn't studied." Fisher knew that infection with HPV could lead to cancer; so she raised the specter that the vaccine could do the same thing. But HPV causes cancer by incorporating two of its genes into cells that line the cervix. Because the HPV vaccine contains no HPV genes, only HPV proteins, it's not possible for the vaccine to do this.

By 2009, more than thirty million doses of HPV vaccine had been given without serious consequences. Women who choose to believe Barbara Loe Fisher's warnings and refuse the HPV vaccine are at increased risk of cervical cancer: an event that occurs twenty to twenty-five years after this very common infection.

Fisher's ill-conceived notion that vaccines cause chronic diseases isn't the only example of how she has squandered an opportunity to alert the media and parents about the real issues of vaccine safety. There were, unfortunately, many others:

- In 1995, when the chickenpox vaccine was recommended for all children, Fisher objected. "Certainly if your child has leukemia or a compromised immune system, you should have them vaccinated. But for your average kid, chickenpox is not a serious disease." Before the vaccine, about ten thousand children were hospitalized and seventy killed by chickenpox every year. Most were previously healthy. Fisher also warned, "the chickenpox vaccine is just going to drive chickenpox into the adult population, where it can be deadly." The chickenpox vaccine, which by 2009 had been used for fifteen years, has led to a 90 percent decrease in the disease, in both children and adults.

- Fisher also decried the concept of herd immunity. "If vaccines are as effective as they are touted to be," she said, "then those who are vaccinated will not face any risk from those who are not." This is clearly untrue, as illustrated by a particularly instructive measles outbreak in the Netherlands between 1999 and 2000. Researchers found that children were actually at greater risk if they were fully vaccinated and living in a relatively unvaccinated community than if they were unvaccinated and living in a highly vaccinated community. That's because no vaccine is 100 percent effective and the greater the likelihood of exposure to a disease, the greater the risk.

- Fisher argued that diseases prevented by vaccines aren't really that bad. "We have gotten into a mind-set where there is an abject fear of disease," she said. "In the '50s, everyone had measles and mumps, and there wasn't this drama attached." Before the vaccine, measles caused more than a hundred thousand hospitalizations and five hundred deaths in the United States every year. Unlike Barbara Loe Fisher, who survived her encounter with measles, those who succumbed aren't around to tell their stories.

- Fisher also argued that natural infection is better than immunization. "Experiencing infectious disease, including influenza, has been part of the human condition since man has walked the earth," she blogged. "Why do vaccinologists insist on assuming that the human immune system is incapable of dealing with that experience? Or benefiting from it? Where is the evidence that it is good to never, ever get the flu?" Before the influenza vaccine was routinely given to children, two hundred thousand were hospitalized and a hundred killed by the virus every year. In 2009, during the H1N1 (swine flu) pandemic, more than a thousand American children died from the flu.

Perhaps Barbara Loe Fisher's most enduring legacy can be found in the comments of Samuel Berkovic, the Melbourne neurologist who had found the real cause of seizures and mental retardation in children claimed to have been damaged by pertussis vaccine. Berkovic remembered the reaction of parents after his discovery: "Most of them were incredibly grateful because they had carried the guilt. They had taken the child to the doctor or maternal/child-health nurse and handed the kid over, and the infant got the vaccine. It was their fault. [They thought] 'if only I'd listened to the lady down the street who said "Don't give your kid a vaccine," I'd have a healthy child.' And when we tell them that that's not right, that sadly your child had this change in the sodium channel that happened at conception and there was nothing you could do about it—that your child was destined to get this—there was an enormous sense of relief. For most, it's relieved decades of guilt."

By telling parents that diabetes, multiple sclerosis, asthma, allergies, seizures, mental retardation, paralysis, and autism are all caused by vaccines, Barbara Loe Fisher—whether intentionally or not—puts the burden of their children's illnesses squarely on the shoulders of those parents. If only they hadn't gotten their children vaccinated, none of these horrors would have happened. But now it's too late; the only thing left is to be angry—angry at a government lost in bureaucracy, pharmaceutical companies in it for the money, and doctors who don't care. By claiming that vaccines cause chronic diseases, Fisher contributes to this cycle of guilt, anger, and blame.

Barbara Loe Fisher is a lightning rod for the media. She is a powerful voice during congressional hearings. And she is the one-stop shop for parents concerned that vaccines are hurting their children. But for all the attention she garners—for all the magazine stories and radio and television interviews—studies have consistently failed to support her concern that vaccines are causing chronic diseases. At the same time that Fisher diverts the media away from real problems with vaccines, one vaccine-safety advocate, working quietly and behind the scenes, has accomplished a great deal. His name is John Salamone.

Salamone was born in upstate New York, in the town of New City. When he was nineteen years old, while studying journalism and government at the University of Maryland, he became the youngest legislative director in Congress. Later, he handled congressional affairs for the Immigration and Naturalization Service. After fifteen years with the government, Salamone moved to the private sector, heading the nonprofit National Italian-American Foundation.

In 1990, Kathy and John Salamone brought their son David to their pediatrician's office in suburban Washington, D.C., to get his vaccines—one of which was Albert Sabin's oral polio vaccine. Two weeks after the visit, the Salamones noticed something wrong with their son. "He couldn't turn over," John said. "He could only move his head back and forth." David Salamone was completely paralyzed below the waist. Eventually he regained the use of his left leg, but his right leg withered. Doctors fitted David with a leg brace and he learned to walk with a gait that, according to John, resembled "a drunken sailor." David's diagnosis was elusive. "The doctors were quite perplexed," recalled Salamone. "They couldn't figure it out because, let's face it, not too many doctors these days are experts in polio. They just simply diagnosed him with a neuropathy of unknown etiology."

David wore the heavy leg brace and started physical therapy. "They noticed that he was in pain," said Salamone, "which was unusual for his condition. And so they sent us to a rheumatologist at Georgetown [who] said, 'Let me just try some tests here.' For the first time, at age three, after being sick quite a bit and always on antibiotics, he was diagnosed with [a congenital immune deficiency]. [Children with severe immune deficiencies are more likely to be paralyzed by the oral polio vaccine.] So now they had figured out why he was sick all of the time with these flu-like symptoms. Then they started to put it all together, to connect the dots and figure out that he got polio from the vaccine."

Salamone learned that a safer vaccine (Salk's), unavailable in the United States, had been used in countries such as Sweden, Norway, and Finland and some provinces in Canada with success, completely eliminating polio. He wanted to know why health officials in the United States used Sabin's vaccine instead of Salk's when the former caused a rare but dangerous side effect and the latter didn't. So,

John Salamone changed polio-vaccine policy in the United States after his son, David, suffered a crippling side effect. (Courtesy of Joseph M. Valenzano, Jr.)

John Salamone became a vaccine-safety advocate, founding an organization called Informed Parents Against Vaccine-Associated Polio (IPAV). "At first you feel guilt," recalled Salamone. "Because you say to yourself, 'we brought our child to the doctor and gave him polio.' Then we got mad, upset, when we found out that there were other options of polio vaccine out there. So we said to ourselves, 'Why wouldn't they be giving that to everyone? Why would you even take a chance?' Then we found out that there were a number of kids every year who were getting polio from the vaccine and I started to identify them little by little and communicate with those families. And in the meantime I wrote to my congressman and I wrote to the White House and I wrote to anyone I could write to and really it was falling on deaf ears. It seemed that if the price was a dozen kids or so a year that got polio from the vaccine well then that was just the price of having a universal vaccine program."

The turning point came in 1993, when Salamone was asked to speak before the Institute of Medicine. He remembered what happened: "The next thing you know an [Associated Press] article picked up on my story and before you knew it I was getting calls from all these newspapers. At one point I was in Italy and some-body called me up and said there's a full page in the *Washington Post* on the front page of the health section with a picture of you and David and the whole polio story."

Salamone knew that if he were to change policy, he was going to have to convince the Advisory Committee on Immunization Practices (ACIP), which advises the CDC. "From there I decided that the only way we were going to make a change was to convince the ACIP and people down at the CDC that there was a need to make this change," recalled Salamone. Every time the ACIP discussed the polio vaccine, John Salamone was there. And every time the AAP held its annual meetings, Salamone was sitting behind a booth manned by several children paralyzed by the vaccine. Salamone converted statistics into real people. Always respectful, clear spoken, and passionate, he was very persuasive. By 1998, the ACIP had switched to the inactivated polio vaccine, eliminating paralysis caused by the live, weakened vaccine. John Salamone's advocacy had everything to do with that change.

In 2008, after twenty-five years with the National Italian-American Foundation, John Salamone retired. He and Kathy now live in Mount Holly, Virginia. "I was lucky," recalls Salamone. "My degree is in journalism and my background is the Congress. If you believe in a God, He picked the right family to have a son who got polio from the vaccine. Because I had the combination of the ability to write and I had good contacts in Washington." Salamone argues that successful vaccine advocacy depends on two things: "I think what worked was organizing families [and] getting the media involved. If the media hadn't picked up on this, I would never have had my opportunity to go before the committee."

In many ways, John Salamone and Barbara Loe Fisher are similar. Both are passionate advocates, both are superb writers and powerful speakers, both have had ready access to the media, and both have had the opportunity to make their case in front of congressional committees. But whereas Salamone's concerns have been supported by science, Fisher's haven't. "[Fisher's group] sends me material and I put it in the circular file," says Salamone. "It seems to me that we can't be guided by bad science. It would be reckless to go out there and attribute certain side effects to a vaccine when they're not documented by science. Attacking vaccines with bad science and bad information is nothing short of reckless and irresponsible."

Ironically, Salamone found that Barbara Loe Fisher's group hurt his efforts to change policy. When Salamone manned a booth at AAP meetings he was often confronted by angry physicians: "Their response, candidly, was not great. My biggest problem was that when you talked about vaccine injury, you were immediately labeled as anti-vaccine. They wanted to put me in the same category as the Fishers of the world, the organizations that had a broader agenda of reducing if not eliminating many of the vaccines. To the contrary, I was pro-vaccine. But I was pro–vaccine safety. I was knowledgeable enough to know the history that many more people's children and adults have been saved by vaccines than have ever died from them."

Although John Salamone and Barbara Loe Fisher are both vaccine-safety activists, they have different core beliefs. First: Barbara Loe Fisher doesn't trust doctors. This distrust was born of a time when she felt that they had lied about—or at least neglected to mention—harmful effects of pertussis vaccine. And she never forgot it. She portrays doctors as elitists and frauds. In *A Shot in the Dark*, referring to doctors, she quoted Shakespeare's *Measure for Measure*:

> But man, proud man,
> Drest in a little brief authority,
> Most ignorant of what he's most assured,
> His glassy essence, like an angry ape,
> Plays such fantastic tricks before high heaven
> As make the angels weep.

As the years passed, Fisher's anger toward doctors, scientists, and public health officials only intensified. She accused them of deception: "Do public health authorities and pediatricians really think that mothers and fathers are going to passively accept the brain-damaging and killing of their children with unsafe vaccines? Do they really believe that the letters written after their names will protect them when the public knows the whole truth about what they have done to so many innocent children around the world?"

She accused them of not caring: "The medical 'experts' are not infallible and they are not immune from the desire for power, money, and fame. The failure to value the life of each individual child by writing off some children as expendable in service to the rest is at the root of much of the distrust educated parents have for the mass vaccination system and those who operate it."

And she likened them to doctors in Nazi Germany: "We would do well to remember what happened in pre–World War II Germany, when doctors were legally allowed to kill anyone in society considered to be a threat to the public health and welfare. Are we preparing for the day when doctors can not only kill handicapped newborns but also handicapped older children and adults, including the vaccine-injured who face a lifetime of long-term care?"

John Salamone had every reason to question doctors. When his son got the oral polio vaccine, he was never told that there was a chance he could suffer permanent paralysis. But Salamone didn't see doctors as evil. Rather, he believed he could convince them that choosing Sabin's vaccine over Salk's was illogical and dangerous. He believed that he could appeal to their ability to reason—that they were human and would respond humanely. And he was right. Salamone's rational advocacy earned the respect of doctors and policy makers. The secretary of health and human services asked him to serve on the Advisory Commission on Childhood Vaccines and, later, the ACIP—the very committee he had implored to switch to a safer polio vaccine.

In contrast, Barbara Loe Fisher, rather than appreciating the researchers who spent millions of dollars on studies to answer her questions, had disdain for them. She believed that studies that failed to support her beliefs were useless—or, worse, fabricated. "Those determined to deny an association between vaccine-induced inflammatory conditions in the body usually like to use retrospective, case-controlled 'studies' that look at old medical records," she blogged. "Using pencils and calculators to dismiss causal associations between vaccines and chronic diseases is easier than having to look at real live patients. . . . The pathetic attempts by the pencil pushers to write off onset of brain and immune dysfunction after vaccination in previously healthy people as just a 'coincidence' will not wash. The people, whose lives have been ruined by doctors too proud to admit harm being done, will not let them get away with it." Fisher's anger wasn't limited to blog postings. In July 2002, at a meeting of parents, public health officials, doctors, and representatives from pharmaceutical companies, she berated the group. One attendee recalled, "On the last day of the meeting, Roger Bernier, the chief architect of the event, was [telling] everyone how important it was that we were all there participating in developing a plan for how we could all work together on vaccine safety issues. He spoke about the importance of the meeting, praised everyone for being there, and said that we were all being pro-active in moving ahead with this plan, before we had a crisis. Fisher was the next to speak. She was fiery. She retorted, 'We *are* in a crisis. And if we don't move forward with this plan, there will be a bloodbath! More government hearings—you bet! More media stories—absolutely!'"

In the early 1980s, Barbara Loe Fisher watched personal-injury lawyers win hundreds of millions of dollars in settlements and awards for alleged victims of pertussis vaccine. It was in a courtroom, not a scientific venue, that Fisher found a sympathetic ear. But with the birth of the National Childhood Vaccine Injury Act in 1986, parents now had to go through a special court. In 2009, this court would be put to the test. Again, vaccine activists and lawyers would seek hundreds of millions of dollars in compensation. It would be their next big bet. This time the complaint wasn't epilepsy or mental retardation or learning disabilities; it was autism. And there was every reason to believe that the success of the pertussis litigation was about to be repeated.

CHAPTER 6

Justice

He handed us fiction after fiction and we printed
them all. Because we found him—entertaining. It's
indefensible. Don't you know that?

—CHUCK LANE, *SHATTERED GLASS*

On February 12, 2009, the Vaccine Injury Compensation Program (VICP) issued a verdict that made international headlines. For several years the VICP had considered five thousand cases brought by parents claiming that vaccines had caused their children's autism. The trial, called the Omnibus Autism Proceeding, was unusual; only once before had vaccine court considered what amounted to a class-action lawsuit. If successful, the action threatened the viability of a program that had once saved vaccines. It was a seminal, defining moment.

And the verdict surprised everyone.

During the twenty years of its existence, the VICP had been generous with its money. Plaintiffs had several ways to win. The program could simply concede that a claim was valid. For example, money was immediately given to any child paralyzed by Albert Sabin's oral polio vaccine. If a suspected injury (such as autism) wasn't considered to be valid, federally appointed judges called "special masters" heard the cases in vaccine court. If the special masters denied the claim, three options remained. The petitioner could appeal to the

United States Court of Claims or, failing that, to the Federal Circuit Court of Appeals; or, failing that, could sue vaccine makers in state courts. Over the years the amount of money available to the VICP, which came from a federal excise tax on every dose of vaccine, got bigger and bigger; by 2009, the program had $2 billion at its disposal and the average award was $900,000.

Vaccine court was an unusual place. When it first started awarding money in 1988, the compensation table was relatively loose and the VICP compensated injuries that weren't caused by vaccines. For example, in 1992 it ruled that DTP had caused a child's epilepsy, even though the child in question had Rett Syndrome, a genetic disorder known to cause seizures. The court also routinely compensated parents whose children died from Sudden Infant Death Syndrome even though studies had shown that SIDS wasn't caused by vaccines. Other awards were even more bizarre. Petitioners successfully claimed that a child's crying or irritability or sleepiness following DTP was evidence of brain damage leading to attention deficit disorder. One lawyer later acknowledged that at least a third of all claims against DTP weren't valid. Probably the best example of the carnival-like atmosphere of vaccine court was the complaint by one petitioner that, following a rabies shot, her dog was "stupider."

In 1995, everything changed. To applause from the medical community, the VICP's vaccine-injury-compensation table was tightened. No longer could parents successfully claim that DTP vaccine caused permanent brain damage or SIDS or genetic disorders. Barbara Loe Fisher, who had helped create the program, was angered by the change. "The federal compensation system that we were told would be 'simple justice for children' has become a cruel joke." Fisher had no need to worry; the procession of inexplicable rulings had only begun.

First was the strange case of Margaret Althen. On March 28, 1997, Althen, forty-nine years old, received tetanus and hepatitis A vaccines. Two weeks later, she suffered headaches and blurred vision. Following a series of medical tests, Althen's doctors diagnosed a disease similar to multiple sclerosis. By the time the Althen case appeared before the VICP, abundant evidence existed that vaccines didn't cause her condition. So, the special master turned down Althen's request for compensation. But judges in the Federal Circuit Court of Appeals disagreed. They argued that plaintiffs' lawyers had to do only three things: propose a theory for how a vaccine could cause a disease, propose a logical sequence of cause and effect, and show that the two events were connected in time. Plaintiffs' lawyers didn't have to provide any epidemiological or biological evidence to prove their case. All they had to do was show that a disease had followed a vaccine and propose a theory for how it could have happened—a very low bar.

Next was the Capizzano ruling. On May 3, 1998, Rose Capizzano received her second dose of hepatitis B vaccine, following which she developed a rash and stiff, painful joints. Later, doctors diagnosed rheumatoid arthritis. Again, evidence refuted the notion that hepatitis B vaccine caused rheumatoid arthritis. So the special master rejected the claim. But, as had been the case with Margaret Althen, the Federal Circuit Court of Appeals overturned the decision. Judges in the Appeals Court noted that Capizzano's doctor believed the vaccine had caused her arthritis, and that was good enough for them. They ruled, "Medical records and medical opinion testimony are favored in vaccine cases, as treating physicians are likely to be in the best position to determine whether a logical sequence of cause and effect show[s] that the vaccination was the reason for the injury." This ruling meant that the treating physician could trump medical consensus, trump the weight of epidemiological and biological evidence, and trump decades of collective clinical experience. All could be cast aside by the opinion of one doctor.

The Althen and Capizzano rulings opened the door for a series of embarrassing decisions. Between 2001 and 2008, the court ruled that hepatitis B and Hib vaccines caused paralysis, MMR caused epilepsy and fibromyalgia (a disorder of unknown cause characterized by muscle pain and fatigue), hepatitis B vaccine caused Guillain-Barré Syndrome, and rubella vaccine caused chronic arthritis, despite abundant evidence showing that they didn't. But no decision was more illogical, more ill-founded, or (for those concerned about the viability of the program) more worrisome than that involving Dorothy Werderitsch.

On November 11, 1992, Werderitsch, a thirty-three-year-old nurse, received her first dose of hepatitis B vaccine; a month later, she received her second. Soon after, she suffered numbness on the left side of her body; then she lost vision in one eye. On February 2, 1993, after a series of medical tests, doctors had a diagnosis: multiple sclerosis. On May 26, 2006, a special master ruled that Dorothy Werderitsch's multiple sclerosis was caused by hepatitis B vaccine.

The notion that hepatitis B vaccine caused multiple sclerosis wasn't supported by the science. For one thing, the viral protein in the vaccine (hepatitis B surface protein) and the viral protein in blood found during natural infection are identical. So a disease caused by the vaccine should also be caused by natural infection. But multiple sclerosis isn't more common in people with hepatitis B virus infections; it's less common. Also, two large epidemiological studies, performed by researchers at Harvard's School of Public Health and McGill University in Montreal—involving tens of thousands of subjects and published in the *New England Journal of Medicine*—found no evidence of an association. When the special master decided that hepatitis B vaccine caused Dorothy Werderitsch's multiple sclerosis, both of these studies had already been published.

The rulings in favor of Althen, Capizzano, and Werderitsch frightened those concerned about what could happen in the vaccine-autism cases.

Perhaps the best person to provide a look behind the curtain of the Vaccine Injury Compensation Program is Lucy Rorke-Adams, a professor of pathology at the Children's Hospital of Philadelphia. Every time a child's death is linked to a vaccine, Rorke-Adams examines the brain. Her unusual role in the VICP is explained by her remarkable story.

Rorke-Adams is the fifth and last daughter of Armenian immigrants who barely escaped Turkish persecution. Although she was born and raised in St. Paul, Minnesota, her parents spoke only Armenian. "When I entered kindergarten, I couldn't speak English and

Lucy Rorke-Adams, a neurologist at the Children's Hospital of Philadelphia, was a key expert in cases of children suspected of suffering or dying from vaccines. (Courtesy of Lucy Rorke-Adams.)

sat on a swing in the doorway between the classroom and cloakroom and cried," she recalled. Between 1947 and 1957, Rorke-Adams earned a bachelor of arts, master of arts in psychology, bachelor of science in medicine, and doctor of medicine, all from the University of Minnesota. After medical school she began an internship in Philadelphia. There she met the scientist who changed the course of her career. "The chairman of pathology at Philadelphia General Hospital was William Ehrich," she recalls. "Dr. Ehrich had been a pupil of Ludwig Aschoff, who had been taught by Frederich Daniel von Recklinghausen, who in turn was educated by Rudolf Virchow, the father of pathology. I regard myself as Virchow's great-great-granddaughter, scientifically speaking!" (All these legendary German pathologists have been immortalized by medical terms bearing their names.)

After more than forty years in the field, Rorke-Adams is recognized as one of the world's foremost pediatric neuropathologists, consulted by colleagues worldwide. When Albert Einstein died, she received a section of his brain to study. It is a testament to the thoroughness and rigor of defense attorneys in vaccine court that they turn so frequently to Rorke-Adams for her expertise.

Rorke-Adams has now evaluated the brains of thirty-three children who died after vaccination or whose unexplained seizures required a brain biopsy. She is still waiting for one to show that vaccines were the cause. "What I did find," she recollects, "is a variety of abnormalities and sundry diseases that could explain what the kids had." Rorke-Adams found that some children had died from malformations, degenerative diseases, vascular disorders, and infectious diseases; others, from accidental smotherings or child abuse. "So, the bottom line," she says, "is that there is no evidence, in terms of scientific evaluation and now pathological evaluation, of anything that one can ascribe to a vaccine." But despite her expertise, and despite the fact that she has supported her evaluations with cogent, well-researched opinions, Rorke-Adams often finds that petitioners prevail. And they prevail behind their principal expert: John Shane, a man who claims to be an expert in neuropathology even though according to Rorke-Adams he has no specific training or standing in the field.

Shane was the chief of pathology at Lehigh Valley, a community hospital north of Philadelphia. "John Shane has testified under oath that because he was responsible for all of the neuropathology at Lehigh Valley Hospital for forty to forty-five years, that he is an expert in neuropathology," says Rorke-Adams. "His knowledge of the infant brain is sketchy at best. For example, he is unable to distinguish inflammatory cells, called lymphocytes, from cells in the brain of a baby that look like lymphocytes, but which are actually primitive nerve cells. In [the case of] one baby he testified that the vaccine had caused encephalitis [inflammation of the brain], when in reality the brain was entirely normal. Unfortunately, the special master hearing the case allowed herself to be convinced of this absurdity." The cells that John Shane had confused with inflammatory cells are called "primitive neuroectodermal cells": a cell type on which Rorke-Adams, unlike Shane, has published extensively.

Rorke-Adams described another example of Shane's questionable expertise in evaluating infant brains. "Shane claimed that the vaccine virus had damaged myelin resulting in a demyelinating disease," she recalled. But Shane was claiming the impossible.

"Myelination of the developing nervous system is far from complete at the time of birth," says Rorke-Adams. "[That's why] human newborns, unlike calves or foals, cannot jump out of the womb and run around the delivery room. Although myelination isn't complete until early childhood, it is sufficiently advanced by 12 to 18 months of age to allow a baby to crawl and start walking. However, at six months of age, certain portions of the brain have little myelin. If there is little or no myelin it isn't possible to have a demyelinating disease. Therefore, demyelinating diseases aren't generally found in children less than one year of age." (Rorke-Adams is in an excellent position to know this, having published the seminal and much-referenced book *Myelination of the Brain in the Newborn*.)

Shane's problems weren't limited to his lack of expertise in neuropathology. A lawyer friend named John Karoly asked Shane to witness a will of Karoly's brother, Peter, written *after* Peter and his wife had died in a plane crash. Karoly wanted part of his brother's multimillion-dollar estate; so Shane witnessed Peter's faked signature. Unfortunately for Shane and Karoly, Peter Karoly already had a will on record. The criminal complaint against John Karoly, Jr., and John Shane was filed in U.S. District Court on September 25, 2008. According to U.S. Attorney Laurie Magid, "The defendants conspired in a fraudulent scheme to forge the wills of Peter Karoly and Lauren Angstadt in order to unlawfully benefit from their tragic deaths. Their actions were not only illegal; they subverted the true intentions of the victims." Six months later, Karoly was charged in a $500,000 tax evasion scheme. In a plea bargain that included renouncing his claim against his late brother's estate, the previous charges against Karoly and Shane were dropped.

The story of John Shane—a professional witness with questionable expertise and integrity—would be repeated in the Omnibus Autism Proceeding.

Among the first to claim that vaccines might cause autism was Barbara Loe Fisher. In *A Shot in the Dark* she wrote, "With the increasing number of vaccinations American babies have been required to use has come increasing numbers of reports of chronic

immune and neurologic disorders . . . including . . . autism." At the
time, Fisher's claim had little traction; few parents carried the banner. But, when a British doctor said it, everything changed.

In 1998, Andrew Wakefield, a surgeon working at the prestigious Royal Free Hospital in London, published a paper in the well-respected medical journal *The Lancet*. The paper, densely titled
"Ileal-Lymphoid-Nodular Hyperplasia, Non-Specific Colitis, and
Pervasive Developmental Disorder in Children," reported the stories of eight children with autism. Wakefield claimed that all the
children had received MMR, all had symptoms of autism soon after,
and all had inflammation of their intestines. In his paper, Wakefield
proposed a series of events: measles vaccine entered the intestines
causing inflammation; once inflamed, the intestines became leaky,
allowing harmful proteins to enter the blood, travel to the brain,
and cause autism.

As a consequence of Wakefield's paper, many parents abandoned
MMR. (It is interesting to note that fears of both pertussis and MMR
vaccine originated in England. "I think our media [have] a lot to do
with it," says David Salisbury, director of England's Department of
Immunization. "[The United States] has basically three newspapers
that cross the country [the *New York Times*, the *Wall Street Journal*,
and *USA Today*] and the rest are very much local newspapers. We
have at least fifteen national-level newspapers. So our journalists are
competing for coverage of a smaller population divided amongst a
much greater number of papers. And I think that leads to more histrionic, more aggressive reporting to seize the audience." Salisbury sees
another connection between the pertussis and MMR scares: "The
younger mothers of children who are being offered MMR were the
daughters of women who had not [given them] the pertussis vaccine.
So the grandmother was not a trivial person in the MMR issue.")

Predictably, outbreaks of measles swept across the United Kingdom and Ireland, causing hundreds of hospitalizations and four
deaths. In the United States, parents of a hundred thousand children
chose not to vaccinate them. The worldwide panic following Wakefield's paper caused researchers to take a closer look. Investigators

Brian Deer (right), an investigative reporter, confronts Andrew Wakefield. Deer almost single-handedly exposed improprieties in Wakefield's research. (Courtesy of AFP/Getty Images.)

found that children with autism were not more likely to have measles vaccine virus in their intestines; and they were not more likely to have intestinal inflammation. Further, no one identified brain-damaging proteins in the bloodstream of children who had received MMR. Finally, twelve separate groups of researchers working in several different countries examined hundreds of thousands of children who had or hadn't received MMR. The risk of autism was the same in both groups. For scientists, these studies ended the concern that MMR caused autism.

Because no one could confirm his work, Wakefield lost credibility among his colleagues. Then he suffered further disgrace. Brian Deer, a journalist working for the *Sunday Times* of London, found that the parents of five of the eight children described in Wakefield's paper were suing pharmaceutical companies, claiming that MMR had caused autism. According to Brian Deer, Wakefield received as much as £440,000 from the Legal Aid Board to act as an expert in the litigation. When Wakefield's co-authors found out about the money, ten of thirteen formally withdrew their names from the paper, distancing themselves from the MMR-causes-autism hypothesis. Finally, one of Wakefield's co-workers, Nicholas Chadwick, testified that Wakefield had published the finding that autistic children had measles vaccine virus genes in their spinal fluid when he knew that the test was incorrect. Wakefield eventually left England, landing at a clinic called Thoughtful House in Austin, Texas, where he offered a variety of "cures" for children with autism.

Throughout Wakefield's rise and fall, Barbara Loe Fisher supported him.

In 2000, after biological and epidemiological studies refuted his hypothesis—and after measles had killed four children in England and Ireland because their parents were afraid of MMR vaccine—Barbara Loe Fisher's National Vaccine Information Center gave Andrew Wakefield the "Courage in Science Award."

In 2002, after Kreesten Madsen and his colleagues published a large, carefully performed study of Danish children in the *New England Journal of Medicine* showing no association between MMR and autism, Fisher refused to believe it: "I can tell you this has not put everything to rest for parents of kids who are functioning perfectly well and then get vaccinated and start to regress. The experience of the people is coming up against the wall of denial by science and medicine."

In 2004, after the Institute of Medicine concluded that evidence clearly refuted Wakefield's hypothesis, Fisher again saw foul play. "This report is a case of political immunology masquerading as real science," she said. "With it, the Institute of Medicine takes a step toward weakening its reputation as an independent body capable of making an objective scientific analysis [that isn't] influenced by government policy and industry profits." As had been the case with diabetes and multiple sclerosis, Barbara Loe Fisher refused to believe that scientific studies exonerating vaccines were anything other than a vast international conspiracy to hide the truth.

In January 2010, after England's General Medical Council ruled that Wakefield had acted with "callous disregard" for children when

Andrew Wakefield leaving the General Medical Council hearing into his alleged misconduct, July 2007. (Courtesy of Lindsey Parnaby/epa/Corbis.)

he had subjected them to spinal taps, endoscopies, and intestinal biopsies; had brought the medical profession "into disrepute" when he had paid children £5 for their blood at his son's birthday party; and had "failed in his duties as a responsible consultant" in not getting approval for his studies from an ethics review board, Barbara Loe Fisher continued to stand by him. "The General Medical Council inquisition was never about the three doctors they put on the rack and found guilty on most counts," she wrote. "It was always about declaring vaccine science and policy innocent on all counts. And creating a horrible warning to any young doctor, who even thinks about investigating or talking about better defining vaccine risks, to think again, shut up, and salute smartly."

One week later, when *The Lancet* formally retracted Wakefield's paper, Fisher fought back. Typically, bad science disappears in a fog of irreproducibility, never requiring a formal retraction. Journal editors retract only those studies they believe were falsified or misrepresented. Fisher, however, saw conspiracy: "[The retraction] sends a signal to the rest of the scientific community that when you dare to investigate the link between vaccination and autism, you do that at your own professional risk." Barbara Loe Fisher failed to note that many scientists before Wakefield had published papers proving the rare tragic consequences of vaccines without risking their career—for example, William Sawyer, who showed that a yellow fever vaccine had been contaminated with hepatitis B virus; Neil Nathanson, who showed that a polio vaccine wasn't properly inactivated, causing children to become paralyzed and die; and Trudy Murphy, who showed that an early rotavirus vaccine caused intestinal blockage, killing one child. Scientists and public health officials didn't marginalize Wakefield because he had challenged the belief that vaccines are absolutely safe; they did it because he was wrong—clearly and inescapably wrong.

On February 18, 2010, Andrew Wakefield, tarnished by *The Lancet*'s retraction, resigned his position as scientific director of Thoughtful House. Two months later, the General Medical Council struck Wakefield's name from the medical register, eliminating his ability to practice medicine in England.

One year after Wakefield proposed that MMR caused autism, the vaccine-autism hypothesis shifted. In 1999, the American Academy of Pediatrics and the Centers for Disease Control and Prevention worried that children might be receiving too much mercury in vaccines, called for pharmaceutical companies to remove a mercury-containing preservative called thimerosal. The frightening and precipitous manner in which this was done gave rise to several groups that believed mercury in vaccines caused autism: SafeMinds, Moms Against Mercury, Generation Rescue, and the Autism Action Coalition. Now, parents weren't scared just of MMR; they were scared of any vaccine that contained thimerosal.

Again, the academic and public health communities responded, performing six large epidemiological studies examining the risk of autism in children who had or hadn't received vaccines containing thimerosal. The results were reproducible and clear: thimerosal didn't cause autism. Consistent with these studies, after the spring of 2001, when thimerosal was taken out of all vaccines recommended for young infants, the prevalence of autism continued to climb.

Science had answered the question of whether MMR or thimerosal caused autism. But it would be the special masters in vaccine court who would render the final verdict.

By July 2002, vaccine court, which was designed to handle individual cases, had received more than three hundred claims that vaccines caused autism. It was only the beginning—the number would eventually exceed five thousand. To handle the massive workload, Gary Golkiewicz, who had been the Chief Special Master of the VICP since its inception, decided to let the petitioners' lawyers select a few cases to represent specific theories. Although Golkiewicz's decision made it possible to deal with the onslaught of claims, the Omnibus Autism Proceeding dominated the court's time; of the eight special masters, eight law clerks, and two staff attorneys available to the VICP, half devoted themselves solely to the autism hearings. And Golkiewicz felt the pressure. Although he praised both plaintiff and defense lawyers for their willingness to find "creative solutions to overcome the lack of sufficient resources," he knew the program had been stretched beyond reason. "Dedication and hard work have their limits," he said. "And I sincerely believe that we have reached that limit."

The Omnibus Autism Proceeding entertained two theories. The first was that the combination of MMR plus thimerosal in vaccines caused autism; the second, that thimerosal alone was responsible. Both of these theories would be represented by three cases. The children representing the first theory were Michelle Cedillo, Yates Hazelhurst, and Colten Snyder.

The amount of evidence to be considered was overwhelming. "It should be recognized that the evidentiary record, based upon which I decided this case, is massive," wrote one special master. "This record dwarfs, by far, any evidentiary record in any prior Program case. The record contains about 7,700 pages of Michelle Cedillo's medical records alone. The hearing transcripts totaled 2,917 pages in *Cedillo*, 1,049 pages in *Snyder*, and 570 pages in *Hazelhurst*. In addition, the amount of medical literature filed into the records of the three cases was staggering. I have not attempted to calculate the total number of pages of these documents, but clearly the total runs well into the tens of thousands." The amount of evidence delayed the decision. The three test cases were heard in June, October, and November 2007; but a decision wasn't reached until February 2009.

There were other concerns. If the special masters decided in favor of the five thousand children, the cost could be as high as $4.5 billion, wiping out reserve funds. Although scientific studies didn't support the notion that MMR or thimerosal caused autism, many believed that the special masters would rule in favor of the plaintiffs. For one, the VICP had made a series of recent decisions that weren't supported by science—the oddest of which favored Dorothy Werderitsch's outrageous claim that hepatitis B vaccine caused multiple sclerosis. Also, that "experts" like John Shane had been allowed to testify was bad enough. But special masters had done more than just allow his testimony; they'd deferred to it. Finally, and most worrisome, the program was caught between two competing objectives. Was it the job of the VICP to stick to the science? Or was it to keep dissatisfied parents from suing vaccine makers directly, threatening the availability of vaccines for American children? The court struggled to determine which of these two positions would most likely prevent a repeat of the pertussis-vaccine litigation in the 1980s. Gary Golkiewicz felt that Congress had never clearly defined the program's role. A year before the first verdict in the Omnibus Autism Proceeding, he said, "There is a tension between these two objectives: a tension that affects dramatically the litigation of cases, the parties' arguments, and ultimately who wins."

Despite the fears of many, the special masters made it clear that during the Omnibus Autism Proceeding they were going to stick to the science—plausible theories (as in the Althen ruling) or unsupported opinions from treating doctors (as in the Capizzano ruling) weren't going to be enough. In the case of Colten Snyder, the special master wrote, "No one who observed the hearing could doubt [the parents'] commitment to Colten, or their good-faith belief that Colten's condition [was] the result of his childhood vaccines. In this respect, they mirror the anecdotal accounts of the struggles of many other parents of autistic children. However, in this court, as in all other courts, *subjective belief is insufficient as evidence of causation.*" In the case of Yates Hazelhurst, the special master wrote, "The Hazelhursts' experience as parents of an autistic child . . . has been a very difficult one. [I am] moved as a person and as a parent by the Hazelhursts' account and again extend to the Hazelhursts very sincere sympathy for the challenges they face with Yates. [My] charge, however, *does not permit decision making on the basis of sentiment but rather requires a careful legal analysis of the evidence.*"

On February 12, 2009, the special masters issued their verdicts. They were unanimous. All rejected the notion that MMR plus thimerosal-containing vaccines caused autism, finding not a shred of evidence to support the theory. They rejected the plaintiffs' contention that thimerosal caused immune suppression, that measles vaccine damaged the intestine, and that measles vaccine traveled to the brain and caused autism. Their verdicts were unequivocal, leaving not a crack in the door for successful litigation in the future. One special master wrote, "Sadly, the petitioners in this litigation have been the victims of bad science, conducted to support litigation rather than to advance medical and scientific understanding of autism spectrum disorder. The evidence in support of petitioners' causal theory is *weak, contradictory, and unpersuasive.* This is particularly apparent when considering the impressive body of epidemiologic evidence contradicting their theories." Another likened the plaintiffs' theory to a scene from *Through the Looking Glass*: "To conclude that Colten [Snyder]'s condition was the result of his MMR vaccine, an objective observer would have to emulate Lewis Carroll's White Queen and be able to believe six impossible, or at least highly improbable, things before breakfast."

The special masters also made it clear that the John Shanes of the world—nonexperts posing as experts—would no longer be tolerated, at least not with so much at stake. "The quality of the petitioners' experts paled in comparison to the world-class experts proffered by the respondent," wrote one special master.

The plaintiffs' witnesses suffered withering criticisms:

- Marcel Kinsbourne, a pediatric neurologist, had claimed that measles vaccine virus traveled to the brain and caused autism. The special master noted that Kinsbourne spent most of his time testifying for personal-injury lawyers in court: "Dr. Kinsbourne suffers from the stigma attached to a professional witness—one who derives considerable income from testifying in vaccine [court]. In the twenty years of the Vaccine Program's existence, Dr. Kinsbourne has appeared as an expert witness in at least 185 decisions." Further, the special master noted that Kinsbourne said one thing to his academic colleagues and another in court: "In a book chapter he authored, Dr. Kinsbourne included a chart on the causes of autism. In his testimony, he used the same chart, but with one addition; he included measles as a cause. Dr. Kinsbourne was unwilling to say measles was a cause of autism in a publication for his peers, but was willing to do so in a Vaccine Act proceeding."
- Vera Byers, a Ph.D. immunologist, had claimed that thimerosal in vaccines had damaged the immune system. The special master wrote, "Dr. Byers' credibility was not enhanced by several instances of apparent 'resumé padding.' Her [resumé] indicated that she was still on the faculty at the University of Nottingham, although her work ended there in 2000. Her [resumé] indicated that she was 'Medical Director on the team responsible for filing a [license to the FDA] for Enbrel [an immune modifying drug].' When informed that there was no record at the FDA of Dr. Byers playing any role in the Enbrel licensing application, she stated that the information did not make a difference. Dr. Byers' [resumé] described her as a 'medical toxicologist' with 'hands-on experience in assessing medical damage to over 3,000 patients in the past 15 years.' [But] she had not seen patients, other than in a litigation context, for the prior seven years. Even without considering Dr. Byers' apparent misstatements on her [resumé], I find that she is not a particularly good expert witness. [Her] insistence that it was acceptable to use adult norms to measure the immune function of infants and young children was, frankly, incredible."

- Dr. Arthur Krigsman, a pediatric gastroenterologist, claimed that measles vaccine had damaged the intestine. The special master wrote, "Dr. Krigsman, after working as an attending physician at Lenox Hill Hospital in New York from 2000 to 2004, left that hospital's employ under questionable circumstances. The hospital restricted his privileges to perform endoscopies, an invasive procedure with some risks, in the belief that he was performing medically unwarranted [procedures] for research purposes. Dr. Krigsman later joined Dr. Andrew Wakefield at a medical practice [Thoughtful House] in Texas. In 2005, Dr. Krigsman was fined $5,000 by the Texas Board of Medical Examiners, apparently because the website of the medical practice had represented that he was available to see patients at a time when he was not yet licensed to practice medicine in that state. Dr. Krigsman's [resumé] stated that he was an 'Assistant Clinical Professor' at the New York University Medical School, but Dr. Krigsman acknowledged on cross-examination that he has never taught a class there. His [resumé] also listed four items under the heading of 'Publications.' But on cross-examination Dr. Krigsman acknowledged that only one of the four items had actually been published in the scientific literature." (To put this in perspective, Lucy Rorke-Adams has 266 scientific publications.)

The special masters also revealed a more sinister aspect of the vaccines-cause-autism hypothesis—there was money to be made promoting it. No single expert better represented this unseemly side of the controversy than a Florida-based physician named Jeff Bradstreet, a witness in the trial of Colten Snyder. Bradstreet had promoted several cures for autism including secretin (a hormone made by intestinal cells to stimulate the secretion of pancreatic juices), chelation (to rid the body of mercury), immunoglobulin (administered by mouth and by vein), and prednisolone (a potent steroid that suppresses the immune system). He also prescribed dietary supplements he sold in his office. As noted by the special master, "The nutritional supplements prescribed by Dr. Bradstreet were also sold

by Dr. Bradstreet." Bradstreet had taken care of Colten Snyder for
eight years, during which time Colten had visited his office one hundred and sixty times. Bradstreet had ordered laboratory tests (many
of which were not approved by the FDA); and he had performed several spinal taps and inserted fiberoptic scopes into Colten's stomach
and colon. All these tests and procedures were expensive, potentially
dangerous, and—according to the opinions of expert witnesses—of
no value to the child.

One year after the verdicts on the first theory had been filed, on
March 12, 2010, the special masters ruled on the second theory proposed by the plaintiffs: thimerosal alone caused autism. Again, their
verdict was unequivocal. They ruled that the theory was "scientifically unsupportable," and the evidence "one-sided." And they remained angry at the cottage industry of false hope that had sprung
up around children with autism, writing, "[Many parents] relied
upon practitioners and researchers who peddled hope, not opinions
grounded in science and medicine."

    The special masters had delivered a fatal blow to the vaccines-
cause-autism hypothesis. But who really lost? It certainly wasn't the
plaintiffs' experts. Although singled out for their lack of expertise
and of personal integrity, all will likely testify in future trials, charg-
ing handsomely for their services. And it wasn't doctors like Arthur
Krigsman, referred to by a special master as one of the "physicians
who are guilty, in my view, of gross medical misjudgment." Krigsman will likely continue to offer "cures" for children with autism.
Or doctors like Jeff Bradstreet, who will continue to sell dietary supplements and perform endoscopies and spinal taps on children
whose parents believe in him, and are more than willing to pay out-
of-pocket for his services. And it certainly wasn't the plaintiffs'
lawyers, who will continue to try their cases in vaccine court and
collect their fees independent of whether they win or lose. (The law
firm of Conway, Homer and Chin-Caplan submitted a bill to vaccine court for $2,161,564.01.) Rather, the losers were Theresa and
Michael Cedillo and their daughter Michelle, Rolf and Angela
Hazelhurst and their son Yates, and Kathryn and Joseph Snyder and

their son Colten. Under the misdirection of doctors and lawyers, these parents had come to vaccine court looking for relief from the financial burden of caring for their autistic children. But they'd come to the wrong place. Even setting aside the fact that vaccines don't cause autism, providing services for parents of autistic children through the VICP would have been an ineffective way of getting these children what they needed. Also, it would have unfairly excluded autistic children who hadn't been vaccinated. If doctors and lawyers in the Omnibus Autism Proceeding really cared about the autistic children they represented, they would have directed their time, energy, and money toward finding mechanisms to provide needed services. But they didn't. And that's because the anti-vaccine movement had taken the autism story hostage—scaring parents about vaccines; providing a source of revenue for personal-injury lawyers with direct ties to anti-vaccine activists; and driving patients into the waiting arms of anti-vaccine doctors who sold false hope, usually directly out of their offices. Many in the autism community were angry that anti-vaccine activists had diverted so much attention away from the real cause or causes of autism.

The Omnibus Autism Proceeding showed just how far public health officials and academia had come since the days of *DPT: Vaccine Roulette*. In 1982, when Lea Thompson claimed pertussis vaccine caused brain damage, science hadn't advanced far enough to challenge her contention. During the next twenty-five years, studies showed that children who received pertussis vaccine weren't at greater risk of permanent brain damage and that the real cause of the damage was a defect in how brain cells transported certain elements, such as sodium, across the cell membrane. Unfortunately, by the time scientists figured this out, many companies had permanently abandoned vaccines. The vaccines-cause-autism scare was different; this time the public health community was ready. When the Omnibus Autism Proceeding started, epidemiological and biological studies had already gone a long way toward exonerating vaccines. The science was in, and the special masters knew it. One stated, "The Omnibus Autism Proceeding began in 2003 with a plea by the Petitioners' Steering Committee to 'let the science develop.' The science has developed in the intervening years, but not in the petitioners' favor."

In the 1980s, anti-vaccine crusaders—through a series of devastating lawsuits—drove many vaccine makers out of the business; as a consequence, only a handful remain. This, at least in part, has contributed to two decades of periodic shortages involving almost every vaccine. But, because of the National Childhood Vaccine Injury Act, companies' fears of litigation have almost disappeared. And, although fragile, the market for vaccines has stabilized. Another and far more damaging consequence of anti-vaccine activism has been the activists' passionate campaign against vaccine mandates. They want to make vaccines optional. And, during the past three decades, they've gotten much of what they've wanted.

CHAPTER 7

Past Is Prologue

History, despite its wrenching pain, cannot be unlived;
but, if faced with courage, need not be lived again.
—MAYA ANGELOU

In a sense, Barbara Loe Fisher has been around for a hundred and
fifty years. That's because the anti-vaccine movement didn't start
with the pertussis vaccine in the 1970s. It started with the smallpox
vaccine in the 1850s. Remarkably, every aspect of the modern anti-
vaccine movement—every slogan, message, fear, and consequence—
has its origin in the past. But in contrast to the modern-day
movement, the final outcome of the first anti-vaccine movement is
known—and remains a painful, unheeded lesson in how the rights
of the individual can trump the good of society.

The first vaccine prevented a disease that killed more people than
any other: smallpox. The disease would start benignly with fever,
headache, nausea, and backache—symptoms common to many infectious diseases. The symptoms that would follow, however, were
unmistakable. The face, trunk, and limbs erupted in pus-filled blisters that smelled like rotting flesh—blisters so painful that victims
felt like their skin was on fire. Worse: smallpox was highly contagious, spread easily by coughing, sneezing, or even talking. As a consequence, smallpox infected almost everyone. Pregnant women
suffered miscarriages, young children had stunted growth, many

were permanently blinded, and all were left with horribly disfigur-
ing scars. One in three victims died from the disease.

"The smallpox was always present," wrote a British historian
in 1800, "filling the churchyard with corpses, tormenting with
constant fears all whom it had not yet stricken, leaving on those
whose lives it spared the hideous traces of its power, turning the
babe into a changeling at which the mother shuddered, and making the eyes and cheeks of a betrothed maiden objects of horror to
the lover."

Smallpox killed more people than the Black Death and all the wars
of the twentieth century combined; about five hundred million people died from the disease. And it changed the course of history. The
virus claimed the lives of Queen Mary II of England, King Louis I of
Spain, Tsar Peter II of Russia, Queen Ulrika Eleonora of Sweden, and
King Louis XV of France. In Austria, eleven members of the Haps-
burg dynasty died of smallpox, as did rulers in Japan, Thailand, Sri
Lanka, Ethiopia, and Myanmar. When European settlers brought
smallpox and war to North America, they reduced the native population of seventy million to six hundred thousand. No disease was
more feared, more destructive, or more loathsome than smallpox.

In 1796, Edward Jenner invented a vaccine that eliminated smallpox
from the face of the earth. The idea for how to make it wasn't his.

Jenner was a country doctor working in the hamlet of Berkeley,
Gloucestershire, in southern England. In 1770, a milkmaid noticed
that when she milked cows with blisters on their udders—and suffered the same blisters on her hands—she was protected against
smallpox during the epidemics that periodically swept across the
English countryside. She confided her theory to Jenner, who, after
observing the same phenomenon, decided to test it. On May 14,
1796, Jenner took fluid from a blister of another milkmaid named
Sarah Nelmes. Then he injected it under the skin of James Phipps,
the eight-year-old son of a local laborer. After a few days, Phipps
developed a small, pus-filled blister that eventually fell off. To test
the milkmaid's theory, on July 1, 1796, Jenner injected Phipps with
pus taken from someone with smallpox; Phipps survived. In a let-

Edward Jenner inoculates his young son with smallpox vaccine. (Courtesy of Time & Life Pictures/Getty Images.)

ter to a friend, Jenner wrote, "But now listen to the most delightful part of my story. The boy has since been inoculated for the Smallpox which, as I ventured to predict, produced no effect. I shall now pursue my Experiments with redoubled ardor."

Two years later, in 1798—after many similar experiments—Jenner published his observations in a monograph titled *Inquiry into the Causes and Effects of the Variolae Vaccinae*. (The term *vaccination* is derived from the Latin *vaccinae*, meaning "of the cow.") Jenner's vaccine spread through England with remarkable speed, reaching Leeds, Durham, Chester, York, Hull, Birmingham, Nottingham, Liverpool, Plymouth, Bradford, and Manchester as well as many smaller towns. Within two years, his monograph had been translated into several languages and vaccination had spread to France, Germany, Spain, Austria, Hungary, Scandinavia, and the United States. Universally accepted by all classes, between 1810 and 1820 Jenner's vaccine halved the number of deaths from smallpox.

Then British government officials did something that launched a movement that has never ended. They required vaccination.

The first pleas for compulsory vaccination came from a relatively unknown medical society. In 1850, a group of prominent British physicians founded the Epidemiological Society of London; their goal was to evaluate epidemic diseases under the "light of modern science." Like all physicians in medical societies, they had intellectual interests. But unlike those in other societies, they were also political. Society members wanted to influence the state to take a greater role in public health: "to communicate with the government and legislature on matters connected with the prevention of epidemic disease." And no disease interested them more than smallpox. Society physicians reasoned that the best way to protect the public from the ravages of smallpox was to require vaccination. So they lobbied the British Parliament for a compulsory vaccination act.

On February 15, 1853, the "Bill to Further Extend and Make Compulsory the Practice of Vaccination" was introduced in the House of Lords. The bill required all children to be vaccinated by six months of age; parents who failed to comply could be fined or imprisoned. Passage of the bill was ensured by an outbreak that, from the standpoint of the society's physicians, couldn't have come at a better time. Between 1810 and 1850, deaths from smallpox were in a slow, steady decline. But in 1852—the year before the bill was introduced—smallpox deaths in England and Wales increased from four thousand to seven thousand and in London from five hundred to a thousand. As a consequence, the bill passed easily and parents lined up to vaccinate their children. The rush to vaccinate didn't last long. When parents realized that compulsory vaccination wasn't enforced, vaccination rates fell. As a consequence, the vaccination act of 1853, described by one historian as "a damp squib," had limited impact.

The lesson learned in 1853 resulted in a tougher law. This time the government wasn't going to look away when parents chose not to vaccinate their children. The new act, passed in 1867, clearly defined how vaccination would be enforced and who would do the enforcing. First, medical officers would issue a warning to parents who didn't have a certificate proving vaccination. If the warning was ignored, officers took parents to court, where they faced a stiff fine plus court costs. Because the law targeted the poor—considered less likely to vaccinate their children due to "ignorance and prejudice"—many parents couldn't afford the fines. If parents couldn't or wouldn't pay, their family assets were seized and sold at public auction. And if enough money couldn't be raised from these sales, one of the parents (usually the father) could be jailed for up to two weeks. The compulsory vaccine act of 1867 also overturned a court ruling in 1863 that only one penalty could be levied for failure to vaccinate. Now, the cycle of fines, public auctions, and imprisonment could occur again and again.

Compulsory vaccination spawned the first anti-vaccine movement. In 1866, Richard Butler Gibbs co-founded the Anti-Compulsory Vaccination League (ACVL) with his brother George and his cousin John Gibbs. By 1879, the league had a hundred branches and ten thousand members. By 1900, British citizens had formed another two hundred anti-vaccination leagues. Gibbs urged citizens to protest vaccination as an act of patriotism. "Stay then the hand of the vaccinator," he wrote. "Join us in waging war against a practice fraught with such an amount of disease and death. Let Britannia put her foot on this iniquitous destroying, death-producing interference with nature's laws and crush it out."

Anti-vaccine activists produced hundreds of thousands of handbills, posters, pamphlets, and photographs decrying the horror of vaccination and the motives of those who enforced it. In 1881, they published the *Vaccination Vampire*, which likened doctors to vampires who "hovered over the pregnant woman who waited in the shadow of its wings" and to "raven[s] perched on [pregnant] sheep, waiting to pluck out and devour the eyes of newborn lambs." Other images were even more dramatic. Anti-vaccine activists declared that vaccination "offer[ed] up annually an indefinite number of human sacrifices to propitiate an imaginary Devil," and they compared it to "some savage African tribe that every week sacrificed to an idol two children to guard against smallpox."

The contents of Jenner's vaccine also came under fire. Anti-vaccine activists claimed that it contained the "poison of adders, the blood, entrails, and excretions of bats, toads and suckling whelps" and that it transformed a healthy child into "a scrofulous, idiotic ape, a hideous foul-skinned cripple: a diseased burlesque on mankind." Using gothic images, propagandists distributed pictures of vaccinated children turning into minotaurs, hydra-headed monsters, dragons, the incubus, and Frankenstein.

The rallying point of the movement came at public auctions, where the possessions of those who refused to pay fines were sold. Protests took many forms. In 1889, a Mr. Cockcroft plastered a dresser and clothes wringer with anti-vaccine literature, making them unfit for sale. One protester in Charlbury screwed a table to the floor, claiming that it "grew there and we built the house around it." Because local supporters invariably purchased the furniture and gave it back to the owner, auctions became a joke at the expense of the government.

Public auctions were also a site of violence. In 1887, more than sixty uniformed and plainclothes police fought their way through a mob of angry citizens, broke into the house of a noncompliant resident, and took his furniture. The auctioneer, who was pelted with stones and eggs, required police protection. Auctioneers became increasingly harder to find.

Anti-vaccine rallies took other forms. Mothers staged mock funerals featuring small white coffins symbolizing the death of a child. In 1885, a parade of women marched across London: "There was a brass band playing appropriate music, an open hearse with the child's coffin, a number of mourning coaches filled with women in black, and a banner inscribed 'In Memory of 1,000 Children Who Died This Year Through Vaccination.'" Protesters also paraded in front of the House of Commons, played Chopin's Death March, and carried banners declaring "Murdered by Compulsory Vaccination."

To resist vaccination, mothers hid their children. In 1872, a Leeds woman explained that when the vaccine inspector "comes into the neighborhood, we shut our doors, pull down the blinds, and go upstairs until he's gone; that's how we trick him." One father advised,

"When the vaccination inspector calls round 'seeking whom he may devour,' raise a hue and cry after him, cry shame on him, and both you and your neighbors hoot him out of the neighborhood; drive the wolf from the door and let the authorities know that mothers are mothers still, and that it is a mother's duty to protect her child."

Although more than a hundred and fifty years separate the first anti-vaccine movement from today's, the two share remarkably similar beliefs and practices—some are so striking that it's as if nineteenth-century England has sprung to life in twenty-first-century America.

Doctors are evil: In response to the vaccination act of 1853, John Gibbs (who later co-founded the Anti-Compulsory Vaccination League) wrote *Our Medical Liberties.* "Who could receive with cordiality and respect the Doctor of Physic who should burglariously thunder at the door," he wrote, "armed with scab and lancet, feloniously threatening to assault the inmates therewith, and, no matter how loudly he should protest that he was bent upon a mission of mercy, who could avoid suspecting that his real objects are power and gain."

On November 5, 2006, Barbara Loe Fisher, in an article titled "Doctors Want Power to Kill Disabled Babies," echoed the writings of John Gibbs: "The tragic consequences of allowing one small group of individuals in society—those who choose to become medical doctors (M.D.'s) or scientists (Ph.D.'s)—to make life and death decisions for others [are] that they can become drunk with power and end up exploiting people. Those elitists who would force people to take medical risks or even kill people in the name of the greater good of society cannot and should not be trusted. If the birthing rooms and newborn nurseries of the world become killing fields and those who practice science and medicine become the executioners, then it will be a very short time before nursing homes, doctors' offices, and public-health clinics are legally allowed to stock lethal injections."

Public rallies: Anti-vaccine rallies peppered the English countryside for much of the late 1800s. The most dramatic—and the one that garnered more media attention than any other—took place in Leicester in 1885. Organizers made travel arrangements for a hundred thousand people to attend; as a consequence, it was the largest rally of its kind. Actors entertained protesters by playing "doctors riding cows and holding on by the tail, and mothers at upper windows clasping their infants, while policemen were trying to commit a legal burglary at the keyhole in the street below." The highlight of the show was an effigy of Edward Jenner, hanged, decapitated, and taken to the local police station for arraignment.

The spirit of the Leicester rally—a rally described at the time as "a perfect carnival of public merriment," can be found in today's anti-vaccine protests. In June 2006, organizers staged a rally in front of the CDC in Atlanta. People carried signs and children wore T-shirts bearing anti-vaccine slogans; many dressed in costumes, such as prison outfits. It was like a scene out of Monty Hall's television game show, *Let's Make a Deal*—except for the angry, threatening undertone. Protesters with megaphones screamed epithets at CDC employees as they drove through the crowd on their way to work. Others carried placards with pictures of Walter Orenstein, former director of the CDC's National Immunization Program, and Marie McCormick, chairman of the Institute of Medicine's committee to evaluate vaccine safety. Mimicking the defacement of Edward Jenner's effigy, both images were circled in red, a crude slash across their faces, above the word *TERRORIST* in bold, black lettering.

Paranoia: In the days following passage of the vaccination act of 1853, anti-vaccine activists likened government officials meeting in a late-night parliamentary session to a coven of witches preparing destruction: "In a dark midnight hour, when evil spirits were abroad, when nearly all slept save a few doctors, who were rather awake, whose dictum and nostrum carried the night, this Act was passed, this deed was done. It was a deed worthy of the night, dark as the night."

In 2007, Barbara Loe Fisher described an anti-vaccine rally at a Maryland courthouse: "I talked with a mother hundreds of yards from the front of the Courthouse door. I was about twelve inches inside a row of large cement balls that apparently were erected as a barrier to prevent terrorist attacks. I did not know I wasn't supposed to be talking with this Mom inside the barrier. All of a sudden, out of the corner of my eye I saw an armed guard with a dog emerge from the Courthouse and walk toward us. I got a sick feeling in the pit of my stomach. It was the dread that any citizen of any country in any century has ever felt when an armed guard with a dog starts advancing. As if we were common criminals or terrorists, he yelled and gestured for us to move behind the stones. We moved without a word. And the sick feeling in the pit of my stomach told me we were being shown the power of the State wielded by that armed guard with the dog, just as parents inside the Courthouse were being shown the power of the State wielded by doctors with syringes."

The popular notion of extreme coercion by public health officials took yet another form. Following the vaccination act of 1867, colorful accounts abounded of parents watching helplessly as their children were taken away and vaccinated without their consent. Actually, it never happened. Although the act of 1867 talked tough, in practice—apart from the occasional property auction—it didn't act tough. The system literally paid vaccinators to listen to parents' concerns. Some vaccinators bribed parents with beer, baked goods, money, and medicine. Others were perfectly willing to cater to the parents' desire to have the child vaccinated in the home, instead of having to suffer the indignity of a public-vaccine station. Still, rumors that children were taken away and vaccinated persisted. And they persist today.

On October 14, 1999, Jane Orient, a prominent anti-vaccine spokesperson in Tucson, Arizona, appeared on ABC's *Nightline* with Ted Koppel. Orient had just finished likening vaccines to scientific experiments in Nazi Germany. Koppel responded: "Dr. Orient, you raised before that sort of dramatic analogy to the Nuremberg Laws, in which people were required to undergo medical procedures against their will. It's a horrible analogy, and I'm sure you are aware of just how horrible it is. Do you really think that's an appropriate one?" Orient didn't back down. "Yes, I do," she said, "because I think that the CDC is not being honest with people. They are saying these vaccines are safe and to save the world from hepatitis B you have to be vaccinated. If parents want to refuse consent, they may

be threatened with having their children taken away from them."
Koppel, at a loss for words, turned to Dr. Sam Katz, a professor of
pediatrics and infectious diseases expert at Duke University School
of Medicine. "Address if you would, though, what Dr. Orient said
about if parents refuse," asked Koppel, "[namely] that the child
would be taken away from them, because I've never heard of that."
Katz responded: "There's no such event that's ever been recorded to
my knowledge or that I've ever heard from anyone else. That's just
not true." Orient held her ground. "I have heard of cases," she said.
"Well, I have heard of cases of parents." Pressed again, Orient refused to elaborate.

False claims of vaccine harm: In 1802, James Gillray penned a
cartoon that captured the spirit of the time. Titled "The Cow-Pock
or the Wonderful Effects of the New Inoculation," it featured Edward Jenner standing among a group of people, a needle in hand,
ignoring the horror around him. Jenner's vaccine had apparently
turned people into cows; they bore horns, had snouts, or suffered
small cows growing like tumors out of their mouths, arms, faces,
and ears. Looking at this cartoon two hundred years later, one
would assume that Gillray was merely representing the public's concern about the source and purity of Jenner's vaccine. But he wasn't.
People were actually scared they were going to turn into cows. In
the early 1800s, "those opposed to the new practice of vaccination
had reported terrible side effects such as the ox-faced boy or children who ran about on all fours, bellowed, coughed, and squinted
like cows." In 1890, at a meeting of the British Medical Association, one speaker produced a child whom he claimed was "covered
with horn-like excrescences, which had resulted from vaccination."
In 1891, a father said he resisted vaccination because "it is well
known that the bulls go mad every seven years, and that the cows
make them mad." He reasoned that because cows were used to
make vaccine, "the madhouses [were] full of vaccinated children."
During an epidemic of smallpox in Gloucester in 1895, some parents refused vaccination because they were unwilling to have "a
beast be put into their children" for fear that it would cause them
to "low and browse in the field."

Cartoonist James Gillray depicts the fear of British citizens in 1802 that Jenner's smallpox vaccine could turn people into cows. (Courtesy of Time & Life Pictures/Getty Images.)

Other false notions about vaccines were common. George Gibbs, co-founder of the Anti-Compulsory Vaccination League, claimed that it was "statistically demonstrated that Vaccination causes very many more deaths than, even in the worst of times, result from Small-Pox." The National Anti-Compulsory Vaccination League Occasional Circular reported that a child became "spotted over the whole of the body with black, hairy marks . . . as in the negro"; vaccination, according to some, was turning white children into black children. In Westminster, a father tried to exempt his children from vaccination, claiming that it caused diphtheria. Similarly, in New York City in 1916, residents claimed that smallpox vaccine caused polio. Ironically, both the bacterium that causes diphtheria and the virus that causes polio had already been identified at the time these claims were made.

Today's fears of vaccines are far more sophisticated than those of the past. But their biological underpinnings are about the same.

It makes as much sense to say that the MMR vaccine causes autism as to say that the smallpox vaccine turns children into cows. The only difference is that today's claims are couched in scientific jargon, so they sound better. During the Omnibus Autism Proceeding, fringe doctors and scientists claimed that thimerosal weakened the immune system, allowing measles vaccine virus to damage the intestine. Because the intestine was now leaky, brain-damaging proteins could enter the bloodstream and cause autism. These hypotheses sound perfectly feasible, except that not a single aspect of them is correct.

It isn't hard to make even the most preposterous claim seem reasonable. If anti-vaccine activists from the nineteenth century were alive today they would no doubt provide a far more sophisticated rationale for their smallpox-turns-people-into-cows contention. For instance, they could claim that smallpox vaccine is made from the fluid of cow blisters that invariably contain cow DNA. If one injected cow DNA into children, they might argue, it is possible that in a small group of the genetically susceptible children the DNA could incorporate itself into the nucleus of some cells. Cow DNA, which contains the blueprint for making cows, could then take over the cellular machinery, causing small, but noticeable, cow-like features. Were this even remotely possible—given the number of hamburgers containing cow DNA consumed every year, and the fact that small fragments of ingested DNA probably enter the body—we would all be cows by now.

Vaccines are unnatural: A persistent theme of those who opposed vaccination in the nineteenth century was that "parental, conjugal, and domestic rights [include] the right to be pure and unpolluted." The best way to avoid smallpox, argued a Midland anti-vaccinator, was to keep "the blood pure, the bowels regular, and the skin clean." Pure blood was the key. And smallpox vaccine, taken from the lymph of a cow, only made the blood impure.

The sentiment that vaccines contain blood poisons was also expressed on June 4, 2008, during a "Green Our Vaccines" rally in Washington, D.C., led by celebrity anti-vaccine activists Jenny McCarthy and Jim Carrey. "The ingredients," said McCarthy, "like the frickin' mercury, the ether, the aluminum, the anti-freeze, need to be removed immediately, after we saw the devastating effects [they had] on our children."

Rejection of the germ theory: In 1796, when Edward Jenner showed that fluid taken from the blisters of cows protected against smallpox, many didn't believe it. Their disbelief is understandable. Jenner's observation was pure phenomenology, occurring eighty years before the discovery of germs—he had no way to explain why it worked.

One of the first to successfully advance the germ theory was Robert Koch, a German physician. In 1877, Koch showed that a specific bacterium, now called *Bacillus anthracis*, caused anthrax. Then he discovered the bacteria that caused tuberculosis and cholera. By 1900, researchers had found the cause of more than twenty different infections. Koch's observations allowed researchers to explain why Jenner's vaccine worked: infection with cowpox virus protected against disease caused by human smallpox virus. Anti-vaccine activists refused to believe it. Their disbelief was shared by a group of practitioners who, like them, dismissed the scientific advances of the time.

The medical marketplace in nineteenth-century England was broad and diverse, including disciplines that were based on scientific principles, known as allopathic medicine, and those that weren't, known as heteropathic medicine. Heteropathic practices included hydropathy, which claimed the curative power of bathing; homeopathy, which offered medicines that were so dilute there wasn't a single molecule of the active ingredient remaining; and mesmerism, which argued in part that diseases could be treated with magnets. Heteropaths were angry that allopaths had convinced the government to compel a medical practice they didn't offer, seeing compulsory vaccination as allopathic medicine harnessing authority for financial gain.

Anti-vaccine activists joined forces with heteropaths to vigorously denounce the germ theory. One activist argued, "We are being frightened to death by microbes. It is germs, germs, germs everywhere. Must one give up shaking hands, kissing, eating, and drinking? With all the germs ever-present, it is a wonder that any of us are alive at all." Another prominent anti-vaccine activist said, "This infection scare is a sham, fostered, if not got up originally by doctors as a means of raising their own importance and tightening their grasp on the throat of the [nation]." Perhaps the harshest critic of the germ theory was the president of the National Anti-Vaccination League. In 1893, he wondered whether "infection [by germs was] merely a theoretical bogey, worked to frighten laymen, and diverting attention from the real enemy of the human race: dirt."

Surprisingly, the notion that the germ theory was ill-founded isn't dead. It's alive and well among chiropractors, who often distribute literature warning of the dangers of vaccines and offer a safe haven for parents frightened by them.

Chiropractic traces its roots to a mesmerist in Iowa. In 1895, Daniel D. Palmer claimed to have made a startling discovery. One of his patients, Harvey Lillard, who had been deaf for seventeen years, wasn't responding to Palmer's magnets. Then Palmer noticed a lump on the back of Lillard's neck. "An examination showed a vertebra racked from its normal position," Palmer recalled. "I reasoned that if that vertebra was replaced, the man's hearing should be restored. With this object in view, a half-hour's talk persuaded Mr. Lillard to allow me to replace it. I racked it into position by using the spinous process as a lever and soon the man could hear as before." It was a miracle—a miracle that would have made infinitely more sense if the cochlear nerve, responsible for sending nerve impulses from the ear to the brain, actually passed through the neck. Nevertheless, Palmer was convinced and a new method for treating disease—chiropractic—was born. Based on Palmer's observation, chiropractors believe that diseases are caused by an imbalance of the flow of energy from the brain, which could be cured by manipulating the spine.

At the time of Palmer's observation, Robert Koch and others were well on their way to proving the germ theory of disease. Palmer didn't believe it. Nor did his son Bartlett Joshua (B. J.), who became a dominant figure among his fellow chiropractors, all of whom had trained at Daniel Palmer's school for chiropractors in Davenport. B. J. Palmer eschewed the germ theory, writing, "Chi-

ropractors had found in every disease that is supposed to be contagious, a cause in the spine. If we had one hundred cases of smallpox, I can prove to you where, in one, you will find a subluxation
[misalignment of the spine] and you will find the same condition in
the other ninety-nine. I adjust one and return his functions to normal. There is no contagious disease. There is no infection."

Because chiropractors didn't believe the germ theory of infection,
they didn't believe in vaccination. Why bother? They could simply
treat diseases like smallpox by manipulating the spine. That chiropractors rejected the germ theory at its birth isn't surprising; that some
reject it today—given the impact of vaccines and antibiotics—is.

As of 2010, about a hundred thousand chiropractors were practicing in the United States.

The lure of alternative medicine: In nineteenth-century England,
alternative medicine was attractive because it was much less invasive, much gentler, much more humane, caring. No surgeries, no
harsh medicines, no grim prognoses. And the explanations for how
alternative therapies worked were easy to understand. Water and
magnets treated diseases.

Alternative medicine is attractive today for the same reasons. A
perfect example is autism: a disorder for which mainstream medicine hasn't found a cause or cure. Practitioners of alternative medicine, on the other hand, claim both. Fringe doctors argue that
autism is caused by vaccines and can be treated with hyperbaric
oxygen, anti-fungal medications, and creams that rid the body of
mercury. Good science gets shoved to the side, in part, because it's
hard to understand. For example, in 2009, researchers published a
paper in *Nature*, one of the world's premier scientific journals. They
found that some children with autism spectrum disorder had a defect in genes that made proteins on the surface of brain cells called
neural cell adhesion molecules. The specific proteins, cadherin 9 and
cadherin 10, help brain cells communicate with each other. Unfortunately, conceptualizing how problems with cadherin 9 and 10
could cause autism is much more difficult than laying the blame on
vaccines; even worse, it doesn't offer immediate hope for prevention
or cure. It is in such a setting that alternative medicines thrive.

Fear of medical advances: Although the germ theory explained *why* Jenner's vaccine worked, a hundred years would pass before researchers figured out *how* it worked. In 1891, Elie Metchnikoff, a Russian microbiologist and pathologist, showed that certain cells in the bloodstream could kill germs. He called them white corpuscles, leukocytes, or phagocytes. Anti-vaccine activists refused to believe Metchnikoff's discovery. Walter Hadwin, a pharmaceutical chemist and one of the anti-vaccine movement's greatest orators, mocked Metchnikoff's findings. On November 1, 1907, more than fifteen years after the establishment of the theory of specific immunity and thirty years after proof of the germ theory (and one year before Metchnikoff won the Nobel Prize in Medicine), Hadwin likened Metchnikoff's immune cells to "Thames policemen [that supposedly went] rollicking round, gobbling up the disease germs and thus extinguishing the imaginary source of the disease."

The inability to accept scientific advances, the desire to protect outdated, disproved theories, and the rejection of new technology weren't unique to anti-vaccine activists. Perhaps nineteenth-century England's best example of the fear of science occurred in 1818 with the publication of a book by a twenty-one-year-old author named Mary Shelley: *Frankenstein.* Shelley was inspired in part by the work of the Italian physicist Luigi Galvani, who showed that if he stimulated the nerve of a dead frog with an electrical current, the frog's leg would twitch. In Shelley's book, Dr. Victor Frankenstein uses electricity (in the form of lightning) to bring the dead back to life. But Frankenstein's monster would eventually break free, terrorizing the community. Shelley's message was clear. Science was powerful but dangerous.

Today's fears of new technology are no different. When the human papillomavirus (HPV) vaccine was first made available in the United States in June 2006, anti-vaccine activists targeted it for elimination. This was in part a response to the relatively new method used to make it: recombinant DNA technology. To make HPV vaccine, researchers took the gene responsible for the surface protein of the virus, known as the L1 protein, inserted it into a small circular piece of DNA, known as a plasmid, and put the plasmid inside yeast cells (specifically, common baker's yeast). When the yeast cells reproduced themselves, they also made large quantities of the HPV L1 protein, as instructed by the plasmid inside them. The L1 protein, which then assembles itself into a structure that looks just like the virus, is used as a vaccine. The same process is employed by one vaccine maker to make four different L1 proteins representing four different strains of HPV. This means that the vaccine contained only four viral proteins. (In contrast, Jenner's smallpox vaccine contained at least two hundred different viral proteins, plus the contaminating proteins in cow lymph.)

By using a technology that produced only one critical viral protein under highly sterile conditions, the science of vaccine making had advanced well beyond the days of harvesting cow blisters. Anti-vaccine activists, however, were unimpressed, claiming that the HPV vaccine caused strokes, blood clots, heart attacks, paralysis, seizures, and chronic fatigue syndrome. The notion that a single viral protein could do all this—when the whole natural replicating virus can't do any of it—was illogical. And although we might dismiss anti-vaccine activists in the mid-1800s who claimed that vaccination transformed some healthy children into oxen destined to graze on all fours and go mad, the biological bases of those claims are as logical as claims against HPV vaccine today.

Vaccines are an act against God: Anti-vaccine protesters saw vaccines not only as an act against nature but as an act against God, frequently using biblical references to make their point. "Like the mother of Moses," proclaimed one anti-vaccine activist, "I have 'hid' my little one. Hers was in danger from the execution of a senseless and murderous law; mine now is; but no ark of bulrushes would avail me, and there is no Pharaoh's daughter to interpose." Referring to the slaughter of the newborns under King Herod, activists likened the consequences of compulsory vaccine acts to children being slaughtered by "Herodian decree."

Activists described vaccination as a perversion of the Christian sacraments, which were supposed to secure the safety of children, not put them in harm's way. Vaccination was "unchristian," a type of "devil worship" that transformed a child into an "anti-Christ."

In a pamphlet titled *Jenner or Christ?* the author described vaccination as the "most outrageous blasphemy against God [and] against
Nature." Mary Hume-Rothery, a prominent anti-vaccine activist in
the 1880s, argued that vaccination fulfilled the apocalyptic proph-
esy in Revelations 16:2 that warned, "Foul and evil sores came upon
the men who bore the mark of the beast." To Hume-Rothery, vaccination scars were the mark of the devil.

The spirit of Mary Hume-Rothery is alive today in Debi Vinnedge,
founder of Children of God for Life in Largo, Florida. Vinnedge is
angry that two human cell strains—which could be used to make
vaccines for the next several centuries—were obtained from volun-
tary abortions in the early 1960s. These cell lines are used to make
vaccines for rubella, chickenpox, hepatitis A, and rabies. Vinnedge
refuses to accept a product made using cells from an abortion, an
act worthy of excommunication. "Casually accepting the use of
aborted fetal cell lines in medical treatments has been a blatant disgrace to humanity," she has said, "a despicable sullying of the value
and dignity of human life, and has lent credibility to the gross com-
mercialization of aborted babies, ripped from their mother's womb
so that someone could turn a profit. We must not become slaves to
the Culture of Death. Using aborted babies as products to help
those children fortunate enough to not have their lives snuffed out
pre-birth is akin to the most vile form of cannibalism imaginable.
Yet we are asked to accept it for every polite reason except one that
begs the question: 'What kind of a civilization have we really pro-
gressed to when we can find no better way to protect ourselves than
by using the remains of murdered children?'" Vinnedge has lobbied
the Vatican's Pontifical Academy for Life without success—the Vatican arguing that vaccines using cell lines originally obtained from
elective abortions promote the greater good by preventing life-
threatening infections. (Ironically, because natural rubella infections
during pregnancy led to thousands of spontaneous abortions every
year, the rubella vaccine, like the Catholic Church, has prevented
many abortions.)

Mass-marketing: Anti-vaccine activists in Victorian England took
advantage of an increasingly print-oriented society. They produced

hundreds of different handbills and pamphlets; engaged in letter-writing campaigns to local and national newspapers; distributed several periodicals throughout England and Wales; hung posters in shop-windows to encourage citizens to talk about the dangers of vaccines; and produced grotesque, graphic photographs of children supposedly harmed by vaccination, perhaps the most disturbing of which was a child suffering from cancer of the eye. The editor of the *British Medical Journal*, Ernest Hart, lamented the success of the anti-vaccine message. "[Anti-vaccine activists have] an extremely energetic system of distributing tracts, inflammatory postcards, grotesquely drawn envelopes, and other means of disseminating their views." On the other hand, those who were promoting the value of vaccines didn't offer "an accessible antidote to these productions."

Today's methods of mass communication include national television programs, Web logs, YouTube, and Twitter. Through these outlets anti-vaccine activists have been able to get their message to millions of people quickly and cheaply. And they're much better at it than public health officials, doctors, and scientists. Rahul Parikh, a pediatrician in Walnut Creek, California, offered a lament in 2008 that echoed the words of Ernest Hart a hundred and fifty years earlier. In an editorial titled "Fighting for the Reputation of Vaccines," Parikh wrote, "Anti-vaccine groups are well organized and passionate. They have used popular settings such as *Oprah* and *Larry King Live* to make strong emotional appeals and get parents to think twice about having their children vaccinated. People, logical or not, do not forget this kind of emotional prowess. On the other hand, our medical and scientific experts counter with accurate evidence and citations of studies, which do not resonate with many parents. Dispassionate messages are not sticky. Gut-wrenching stories . . . are. It is time we change."

Although the anti-vaccine movement in the mid-1800s is similar to today's, certain differences are striking.

Rich versus poor: Laws compelling vaccination in nineteenth-century England were directed against the poor. Authorities believed that the working class, because it was less educated, would be more likely to fear vaccines and less likely to get them. As a consequence, resistance to vaccination sprang up in working-class neighborhoods in East and South London as well as in heavily industrialized towns such as Manchester, Sheffield, and Liverpool. Resisters were primarily journeymen laborers, artisans, factory workers, and small shopkeepers—groups most likely to suffer the public humiliation of vaccine stations.

Today, on the other hand, resistance to vaccines is found in the upper-middle class among parents who are college- and graduate-school-educated, likely to use the Internet to make healthcare decisions, and fully believing that they, too, can become experts in an information age. The problem, however, lies in how they obtain their expertise. Magazine and newspaper articles and the Internet often provide information that is misleading and unnecessarily frightening. And it's not hard to find like-minded people on the Internet, no matter how small the group or how outlandish the belief.

Lawyers: Unlike anti-vaccine protesters in Victorian England, some of today's anti-vaccine activists find a pot of gold at the end of the injury-compensation-program rainbow. Fear of pertussis vaccine in the 1980s led to millions of dollars in awards and settlements. As a consequence, anti-vaccine organizations now work hand-in-glove with personal-injury lawyers, many of whom sit on their advisory boards and help them prepare pamphlets that warn of the dangers of vaccines and describe how to collect money. For example, a press release by Barbara Loe Fisher's National Vaccine Information Center quotes Michael Kerensky, a lawyer from one of the most powerful product-liability law firms in the United States. At the end of the press release is this statement: "In cooperation with the National Vaccine Information Center, Kerensky has developed an educational pamphlet about the National Vaccine Compensation Fund. To request a copy call 1-800-245-0249." Fisher's Web site provides direct links to sixteen personal-injury law firms.

Marketing strategy: Protesters in nineteenth-century England had no trouble labeling themselves anti-vaccine. Indeed, most organized anti-vaccine groups included the word *anti-vaccination* in their names. Today, however, anti-vaccine activists go out of their way to claim that they are not anti-vaccine; they're pro-vaccine. They just want vaccines to be safer. This is a much softer, less radical, more tolerable message, allowing them greater access to the media. However, because anti-vaccine activists today define *safe* as free from side effects such as autism, learning disabilities, attention deficit disorder, multiple sclerosis, diabetes, strokes, heart attacks, and blood clots—conditions that aren't caused by vaccines—safer vaccines, using their definition, can never be made.

In 1898, the British government finally gave in, appeasing angry citizens by passing a conscientious-objection law. People who didn't want to get a vaccine didn't have to. (The term *conscientious objector*, born of England's anti-vaccine movement, was later applied to those who refused to fight in World War I and subsequent wars.) Within a year, the government issued more than two hundred thousand certificates of conscientious objection. By the late 1890s, vaccination rates had plummeted. In Leicester 80 percent of babies were unvaccinated; in Bedfordshire, 79 percent; in Northamptonshire, 69 percent; in Nottinghamshire, 50 percent; and in Derbyshire, 48 percent. Anti-vaccine forces in England had won the day. In Ireland and Scotland, on the other hand, no such movement existed. No anti-vaccine groups were formed, no anti-vaccine pamphlets were produced, and citizens readily accepted vaccination. While vaccination rates in England fell, those in Scotland and Ireland rose. As a result, England became Europe's epicenter of smallpox disease and death.

For anti-vaccine activists in England, the freedom to choose had become the freedom to die from that choice. As in nineteenth-century England, the battle to eliminate vaccine mandates in twenty-first-century America would also be fought in legislatures and courtrooms. And the results would be all too similar.

CHAPTER 8

Tragedy of the Commons

Freedom is the recognition of necessity.
—GEORG WILHELM FRIEDRICH HEGEL

Parents in nineteenth-century England argued that vaccines were impure or unsafe or an act against nature or God. But their anger wasn't directed at doctors so much as at government officials who had no right to tell them what to do—no right to tell them what should be injected into their children. To protesters, compulsory vaccination was an intolerable violation of their civil liberties.

In America, the response to state-mandated vaccines was no different. Indeed, one citizen's fight went all the way to the Supreme Court. The ruling in that case—called "the most important Supreme Court case in the history of American public health"—has been cited in seventy Supreme Court verdicts and, for more than a century, has determined whether states can force parents to vaccinate their children.

It started with a Lutheran minister in Massachusetts.

In May 1899, a case of smallpox occurred in the city of Swampscott, twelve miles outside Boston. By summer, several more cases appeared in Everett and Charlestown, just across the Charles River. By 1901, more than two hundred Bostonians had fallen victim to the disease. In response, the Cambridge Board of Health proclaimed: "Whereas, smallpox has been prevalent in this city of Cambridge and has continued to increase; and whereas it is necessary for the speedy extermination of the disease; be it ordered that all inhabitants of the city be vaccinated." Citizens who refused were fined $5.00. By early 1902, more than 485,000 had been vaccinated. The *Boston Daily Globe* declared, "There is a greater demand for vaccination in Boston than there is for salvation, even though both are free." In 1903, when the epidemic ended, smallpox had infected sixteen hundred people and killed almost three hundred. If city health officials hadn't acted, the toll would have been much greater.

Not everyone embraced the city's mandate. On March 15, 1902, Dr. Edwin Spencer visited the home of Henning Jacobson and offered to vaccinate him. Jacobson refused—then he refused to pay the fine.

Henning Jacobson was born in Sweden in 1856, coming to the United States when he was thirteen years old. In 1882, while a student at a Lutheran college in Minnesota, Jacobson married Hattie Alexander and together they had five children. In 1893, the Church of Sweden Mission Board asked him to found a Lutheran church in Cambridge, Massachusetts. A pious man, charismatic orator, and community organizer, Jacobson was devoted to his congregants; his fundamental belief that God would protect him was at the heart of his refusal to vaccinate.

In July 1902, Jacobson was tried before a Middlesex county district court. After a jury found him guilty, Jacobson appealed to the county's superior court, which, in February 1903, upheld the conviction. Undeterred, Jacobson appealed to the state supreme court. This time, two well-known lawyers represented him: Henry Ballard of Vermont and James Pickering, a Harvard-trained lawyer who would later win fame as the oldest U.S. soldier in World War I. Given their fees and his meager pastor's salary, Jacobson's choice of Ballard and Pickering was surprising. But Pickering lived only a few blocks from Immanuel Pfeiffer, a leader of Boston's Anti-Vaccination League. It was Pfeiffer who had convinced Pickering to take the case.

Pickering and Ballard argued that the state had violated Jacobson's civil rights: "Can the free citizen of Massachusetts, who is not

Henning Jacobson was at the center of a landmark U.S. Supreme Court ruling on the right of citizens to resist vaccination. (Courtesy of the Evangelical Lutheran Church in America Archives.)

yet a pagan, not an idolator, be compelled to undergo this rite and to participate in this new—no, revised—form of worship of the Sacred Cow?" Again, Jacobson lost. So he took his case to the highest court in the land. On June 29, 1903, the United States Supreme Court added *Jacobson v. Massachusetts* to its docket. This time Jacobson chose George Williams, a former congressman from Massachusetts, to represent him. Williams argued that Massachusetts, by requiring vaccination, had violated the Fourteenth Amendment—specifically, that no state could deprive any person of life, liberty, or property without due process of law. Williams's legal brief stated, "A compulsory vaccination law is unreasonable, arbitrary, and oppressive, and, therefore, hostile to the inherent right of every freeman to care for his own body and health in such a way as to him seems best." Williams also argued that vaccines were unconscionably dangerous: "We have on our statute book a law that compels . . . a man to offer up his body to pollution and filth and disease; that compels him to submit to a barbarous ceremonial of blood-poisoning, and virtually to say to a sick calf, 'Thou art my savior; in thee I do trust.'"

On February 20, 1905, the Supreme Court—by a vote of 7 to 2—ruled that the right to refuse vaccination wasn't guaranteed by the U.S. Constitution. Writing for the majority, Justice John Marshall Harlan argued that, in the arena of public health, societal good trumped individual freedom: "The liberty secured by the Constitution of the United States to every person within its jurisdiction does not import an absolute right to each person to be, at all times and in all circumstances, wholly freed from restraint. There are manifold restraints to which every person is necessarily subject for the common good. Society based on the rule that each one is a law unto himself would soon be confronted with disorder and anarchy." Harlan described the Cambridge Board of Health law compelling vaccination as a "fundamental principle of the social compact that the whole people covenants with each citizen, and each citizen with the whole people."

Jacobson v. Massachusetts wasn't the only time the Supreme Court considered a state's right to mandate vaccines. Seventeen years later, in 1922, officials from Brackenridge High School in San Antonio, Texas, expelled fifteen-year-old Rosalyn Zucht because her parents refused to vaccinate her. Unlike Boston in the early 1900s, San Antonio wasn't in the midst of a smallpox epidemic. But that didn't matter. In a unanimous decision the court ruled that Rosalyn's expulsion didn't violate her constitutional rights.

The rulings in *Jacobson* and *Zucht* gave states the right to enforce vaccination. But they didn't set limits on how far public health officials could go. In some states, for example, parents could be criminally prosecuted. But the event that probably worried anti-vaccine activists more than any other occurred in New York City in 1906. One health officer involved in the case later said, "There is very little that a Board of Health cannot do in the way of interfering with personal and property rights for the protection of public health." The officer was referring to the strange case of Mary Mallon.

Mary Mallon was born in Ireland on September 23, 1869, immigrating as a teenager to the United States, where she became a cook for wealthy New Yorkers. In 1906, Mallon was working on Oyster Point, Long Island, for a New York City banker named Charles Warren. Warren had rented the house on Oyster Point from George Thompson. That summer, six people in Warren's family were struck down by typhoid fever.

At the turn of the twentieth century, typhoid fever was a well-recognized problem, infecting as many as thirty-five thousand Americans every year. The disease is caused by the bacterium *Salmonella typhi*, which at that time regularly contaminated food and water. Typically, victims had fever, headache, and malaise followed by chills, loss of appetite, and a rash on the chest and abdomen. After the rash disappeared, people often got sicker, with severe cramping and weight loss and, in some cases, decreased blood pressure and shock. One in ten died from the disease.

When Thompson found that half of Warren's family had contracted typhoid fever, he hired George Soper, a sanitary engineer, to find the source. Soper did his homework, uncovering a series of outbreaks similar to the one on Oyster Point. He found one case in 1900 on Long Island; another in 1901 in New York City; seven in 1902 in Dark Harbor, Maine; four in 1904 among servants in Sands Point, New York; and three in 1906: one in Tuxedo Park, New York, and two on Park Avenue in New York City. All these outbreaks had one thing in common: Mary Mallon was the cook. Twenty-two people had been infected; two had died. When Soper was investigating the outbreaks, *Salmonella typhi* was known to contaminate food. But when cooks spread the disease, they were always sick as well. Mallon, on the other hand, had no symptoms of the illness. Indeed, Mary Mallon was the first person in North America found to be a healthy carrier of *Salmonella typhi*.

After Soper identified Mallon as the likely source of the Warren family outbreak, he wanted to prove it. "I had my first talk with Mary in the kitchen," he recalled. "I was as diplomatic as possible, but I had to say I suspected her of making people sick and that I wanted specimens of her urine, feces, and blood. It did not take

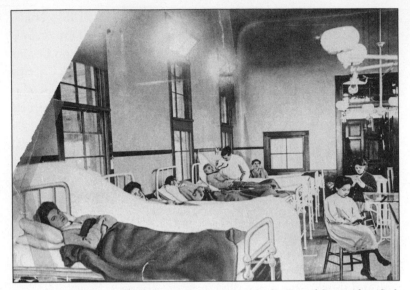

Mary Mallon (left), "Typhoid Mary," the first person in the United States identified as a carrier of typhoid bacilli, was confined on North Brother Island from 1907 to 1910 and then from 1915 until her death in 1938. (Courtesy of Bettmann/Corbis.)

Mary long to react to this suggestion. She seized a carving fork and advanced in my direction." Eventually, Soper gave up, leaving the task of collecting specimens to Josephine Baker, a New York City public health official. On March 19, 1907, Baker, accompanied by a police officer, visited Mary Mallon. "She came out fighting and swearing," recalled Baker, "both of which she could do with appalling efficiency and vigor. There was nothing I could do but take her with us. The policeman lifted her into the ambulance and I literally sat on her all the way to the hospital; it was like being in a cage with an angry lion." Baker took Mallon to the Willard Parker Hospital, then to New York's receiving unit for contagious diseases. There, microbiologists found that her feces were teeming with typhoid bacilli. Immediately, Baker sent Mallon to North Brother Island in the East River, where she lived in a small bungalow, alone, cooking for herself. Mallon didn't understand her incarceration. "I never had typhoid in my life," she said, "and have always been healthy. Why should I be banished like a leper and compelled to live in solitary confinement with only a dog for a companion?" Three years later, health authorities released Mallon after she promised she would no longer work as a cook. The press dubbed her "Typhoid Mary."

In 1915, an outbreak of typhoid fever occurred at Sloane Maternity Hospital in New York City. Twenty-five doctors, nurses, and hospital staff were infected and two died. Investigators traced the source of the epidemic to Mrs. Brown, a cook who had been employed three months earlier. It didn't take long for health officials to realize that Mrs. Brown was Mary Mallon. "Whatever rights she once possessed as the innocent victim of an infected condition were now lost," said George Soper. "She was now a woman who could not claim innocence. She was known willfully and deliberately to have taken desperate chances with human life. She had abused her privilege; she had broken her parole. She was a dangerous character and must be treated accordingly." On March 26, 1915, state health officials again sent Mary Mallon to North Brother Island, where she remained for the rest of her life, dying of a stroke on November 11, 1938. Mallon had spent twenty-six of her sixty-nine years quarantined as a carrier of *Salmonella typhi*, never having exhibited a single symptom of the disease.

American anti-vaccine activists watched what had happened to Mary Mallon—and it scared them. They wondered just how far public health officials would go in their quest to vaccinate children.

Following the Supreme Court's decision in *Jacobson v. Massachusetts*, an editorial appeared in the *New York Times*: "The contention that compulsory vaccination is an infraction of personal liberty and an unconstitutional interference with the right of the individual to have the smallpox if he wants it, and to communicate it to others, has been ended. [This] should end the useful life of the societies of cranks formed to resist the operation of laws relative to vaccination. Their occupation is gone." The prediction couldn't have been more wrong. Anti-vaccine activism in the United States was just getting started.

In 1894, in response to a smallpox outbreak in Brooklyn, the Board of Health ordered Charles McCauley, his wife, and son to be vaccinated. They refused, the elder McCauley threatening the visiting doctor with a rifle. In response, the McCauleys were quarantined; two policemen stood at their door. The *New York Times* reported the incident: "They were forbidden to leave the apartment, and the other tenants were warned, under penalty of arrest, not to deliver any messages to them. The grocers, butchers, and bakers in the vicinity were also forbidden to deliver provisions." The next day, police found a two-foot hole in a closet through which the family had escaped, eventually making it to Hoboken, New Jersey. Three days later, after public health officers convinced the McCauleys they had nothing to fear from the procedure, they surrendered to the Brooklyn police and were vaccinated.

In the late 1890s, incidents similar to that of the McCauleys spawned anti-vaccine groups like the Brooklyn Compulsory Anti-Vaccination League and the Massachusetts Compulsory Anti-Vaccination Association. Local leagues became national organizations. In 1908, two wealthy businessmen, John Pitcairn and Charles Higgins, founded the Anti-Vaccination League of America, a coalition of many smaller groups. Pitcairn declared, "We have repudiated religious tyranny; we have rejected political tyranny; shall we now submit to medical tyranny?" Pitcairn was the orator; Higgins the pamphleteer, producing *Open Your Eyes Wide!* (1912), *The Crime Against the School Child* (1915), *Vaccinations and Lockjaw: The Assassins of the Blood* (1916), and *Horrors of Vaccination Exposed and Illustrated* (1920). One of the most active anti-vaccine groups was the Citizens' Medical Reference Bureau. Founded in New York City in 1919, the group produced the popular pamphlet *The Facts Against Compulsory Vaccination: It's the School and Not the Child That Is Public.*

But of all the anti-vaccine activists at the turn of the century, the most vocal, most passionate, and most irascible was Lora Little, founder of the American Medical Liberty League. Little was moved to action by the death of her son, Kenneth, following a smallpox vaccine. Although the child had actually died from measles and diphtheria, Little was convinced that vaccination was the cause. Like Barbara Loe Fisher, Lora Little believed that doctors and health

The Raggedy Ann doll was created in part to represent children permanently harmed by vaccines. (Courtesy of Lambert/Getty Images.)

officials were all part of a conspiracy to sell vaccines. In her pamphlet, *Crimes of the Cowpox Ring*, Little sounded themes of anti-vaccine activists past and future: "The salaries of the public health officials in this country reach the sum of $14,000,000 annually. One important function of the health boards is vaccination. Without smallpox scares their trade would languish. Thousands of doctors in private practice are also beneficiaries in 'scare' times. And lastly, the vaccine 'farmers' represent a capital of $20,000,000 invested in their foul business."

Anti-vaccine activity in America also produced a popular icon. In 1915, Johnny Gruelle, a cartoonist and illustrator in New York City, watched his daughter, Marcella, die following a smallpox vaccine. Even though the medical report stated that the child had died from a heart defect, Gruelle blamed the vaccine. In his daughter's memory, he created a doll with red yarn for hair and floppy arms and legs—a symbol of children harmed by vaccines. He called it Raggedy Ann.

By the 1930s, as the need for smallpox vaccine declined, so did the passion of anti-vaccine groups. The Anti-Vaccination League of America lost its voice when Charles Higgins died in 1929, as did the American Medical Liberty League when Lora Little died two years later. Massive programs promoting diphtheria vaccine in the 1930s and polio vaccine in the 1950s failed to incite anti-vaccine activity. It would be another fifty years, until the airing of *DPT: Vaccine Roulette*, before the nationwide anger spurred by state-enforced vaccination reemerged. But the fight against compulsory vaccination in court was far from over.

In the late 1960s and early 1970s, public health officials at the CDC decided to eliminate measles from the United States. Their method sparked a series of lawsuits.

In the first half of the twentieth century, measles caused thousands of hospitalizations and hundreds of deaths every year. With the invention of a vaccine in 1963—and buoyed by the virtual elimination of smallpox and polio—the CDC saw an opportunity. Because measles occurred primarily in school-aged children, health officials reasoned that the best way to eliminate it would be to mandate vaccine for school entry. A famous photograph at the time—and one shown on *Vaccine Roulette*—was that of the CDC's Alan Hinman holding up a sign that read "No Shots, No School."

The CDC picked an opportune moment to launch its attack. Although measles vaccine had caused a dramatic decline in the disease, immunization rates had become stagnant. And measles cases increased. In 1970, forty-seven thousand cases were reported. By 1971, the number had risen to seventy-five thousand.

In addition to state health officials, CDC efforts drew on two powerful political groups. The first was the Joseph P. Kennedy Foundation, the second was that of Betty Bumpers. Kennedy was the father of President John F. Kennedy and senators Robert and Edward Kennedy. He was also the father of Rosemary Kennedy. Severely retarded from birth, Rosemary was lobotomized in 1941 and

it's the law
No Shots -
No School

Bumper sticker displayed by South Carolina public health officials in 1980.
(Courtesy of Dr. Alan Hinman.)

institutionalized from 1949 until her death in 2005. Moved by her struggle, the Kennedy Foundation provided services for the mentally disabled. The foundation was interested in measles because the virus caused encephalitis in four thousand Americans a year, often resulting in permanent brain damage. At the time of the CDC initiative, Edward Kennedy was a senator from Massachusetts and his sister, Eunice Kennedy Shriver, was head of the U.S. Office of Economic Opportunity, created as part of Lyndon Johnson's Great Society. Writing to governors and congressmen, the Kennedys used their considerable influence to promote the enactment and enforcement of local school-entry mandates.

Whereas the Kennedys provided an emotional impetus for school requirements, Betty Bumpers, wife of Arkansas senator Dale Bumpers, provided a financial one. Recruiting future First Lady Rosalynn Carter to her cause, in the early 1970s Bumpers helped create a federally funded program called the Childhood Immunization Initiative that bought vaccines for children whose parents couldn't afford them. Through her efforts, federal funding for vaccines increased from $5 million in 1975 to $17 million in 1976 to $23 million in 1977. In addition, in response to a deadly measles epidemic in the early 1990s, Betty Bumpers and Rosalynn Carter founded Every Child by Two (ECBT), a nonprofit organization. ECBT organizes educational programs in cities with low immunization rates and supports efforts to vaccinate children in Africa.

One result of these efforts was that the number of states requiring vaccines for school entry increased from twenty-five in 1968 to forty in 1974—a situation that provided a natural experiment to see whether school mandates worked. At first, they didn't seem to: outbreaks of measles continued. The problem, it turned out, wasn't that health officials didn't have school-entry requirements; it was that they didn't enforce them. But the tide was about to turn.

In 1976, during a massive measles outbreak in Alaska, state health officials informed parents that their children couldn't come to school until they were immunized. Fifty days later, after more than seventy-four hundred students hadn't complied, health officials excluded them from school. Within a month, fewer than fifty children remained unvaccinated and the epidemic ended.

In 1977, a measles epidemic occurred in Los Angeles County; thousands of children were infected, many suffered measles pneumonia, three had measles encephalitis, and two died. On March 31, the county health director declared that any child who had not had a measles vaccine by May 2 would not be allowed in school. When the deadline arrived, tens of thousands of students still hadn't been vaccinated. As in Alaska, Los Angeles County parents soon realized that health officials weren't kidding—fifty thousand children were barred from school. Within days, most were back with proof of immunity, again ending the epidemic.

Although events in Alaska and Los Angeles County were compelling, the most dramatic example of the power of school mandates occurred in Texarkana, a city that straddles the border between Texas and Arkansas. Between June 1970 and January 1971 Texarkana suffered 633 cases of measles. At the time of the outbreak, Arkansas required vaccines for school entry and Texas didn't. Of the 633 cases, 608 (96 percent) occurred in Bowie County (the Texas side) and 25 (4 percent) occurred in Miller County (the Arkansas side).

By 1981, all fifty states had school immunization requirements. During the next two decades, with greater enforcement, the number of measles cases declined dramatically. In 1998, only eighty-nine cases of measles occurred in the United States, most imported from other countries.

Walter Orenstein was a young epidemiologist at the CDC during the 1977 measles outbreak in Los Angeles: "The school mandate [in Los Angeles] really changed everything," he recalled. "That set the precedent for no shots, no school. The beauty of the school laws is that we didn't have policemen forcing people against their will to be vaccinated. We just said if you are unwilling to be vaccinated, you can't go to school." Orenstein was drawing a legal distinction between compulsory vaccination, in which people who refuse are forcibly vaccinated, and mandatory vaccination, in which people who refuse are denied social privileges, like attending public school.

Despite the remarkable success of school mandates, the heavy hand of health officials provoked a backlash. Some parents couldn't accept that their unvaccinated children could be denied an education. So they sued the states, claiming protection under the First Amendment, which requires that "Congress shall make no law respecting an establishment of religion or prohibiting the free exercise thereof. . . ." The ruling that tipped the balance in these First Amendment cases wasn't the Supreme Court ruling in *Jacobson* or *Zucht*. It was one that on its surface had nothing to do with vaccines.

On December 18, 1941, Sarah Prince, a Jehovah's Witness, took her two sons and nine-year-old niece, Betty Simmons, "to engage in the preaching work with her upon the sidewalks." Prince had been warned by a school-attendance officer that bringing children with her at night violated state labor laws, but she ignored the warning. At 8:45 p.m., "Betty held up in her hand, for passers-by to see, copies of 'Watchtower' and 'Consolation.' From her shoulder hung the usual canvas magazine bag." The truant officer again warned Prince to get the children off the street. "Neither you nor anyone else can stop me," threatened Prince. "The child is exercising her God-given right and her constitutional right to preach the gospel, and no creature has a right to interfere with God's commands." The case, known as *Prince v. Massachusetts*, reached the U.S. Supreme Court in 1943, and in January 1944 the court ruled against Prince, arguing that religious freedom didn't trump child-labor laws. Justice Wiley B. Rutledge, who wrote the majority opinion for the

court, then went beyond the case in front of him: "The right to practice religion freely does not include liberty to expose the community
or the child to communicable disease or the latter to ill health or
death. *Parents may be free to become martyrs themselves, but it
does not follow that they are free, in identical circumstances, to
make martyrs of their children.*"

The ruling had a major impact on vaccine lawsuits. In 1965,
the Wrights, members of the General Assembly and Church of the
First Born, argued unsuccessfully that DeWitt High School in
Arkansas County, Arkansas, had violated their right to exercise
their religion by requiring a smallpox vaccine for their children.
In 1968, Thomas McCartney, a chiropractor and Roman Catholic,
argued unsuccessfully that school officials in Broome County, New
York, had violated his religious freedom by requiring a polio vaccine for his ten-year-old son. In 1974, Ronald Avard argued unsuccessfully that the public school system in Manchester, New
Hampshire, had violated his religious freedom by dismissing his
unvaccinated six-year-old son from kindergarten. In 1979, Charles
Brown, a chiropractor and member of the Church of Christ, argued unsuccessfully that the school board in Houston, Mississippi,
had violated his religious freedom by dismissing his unvaccinated
six-year-old son from first grade. In 1982, Irving Davis, a member
of the Church of Human Life Science, argued unsuccessfully that
school officials in Cecil County, Maryland, had violated his religious freedom by not allowing his unvaccinated eight-year-old son
to enter school.

The judges in each of these cases ruled that the right to practice
religion freely hadn't been violated because vaccination was required of all religions, without discrimination. However, a little-
publicized bill in New York State had already opened a crack in the
door—a crack that would have a major impact on the spread of
contagious diseases in the United States. On June 20, 1966, New
York State assemblymen voted on a bill requiring polio vaccine for
school entry. There was one catch. The bill excluded children whose
parents' religion forbade vaccination, a direct result of lobbying by
what was then one of the most powerful religious groups in the

United States: Christian Scientists. The bill passed by a vote of 150 to 2. One of the dissenters, Joseph Margiotta, later explained his reasoning. "Suppose an exempted child was a polio carrier," he said. Margiotta's concern proved prescient.

Founded in 1879 by Mary Baker Eddy, Christian Scientists believed that illness was a mental, not physical, disorder. In her book *Science and Health*, Eddy wrote, "We have smallpox because others have it, but mental mind, not matter, contains and carries the infection." Diseases like smallpox could be avoided by prayer, not by vaccines. Those who chose to follow Eddy's teachings have paid the price.

In 1982, nine-year-old Debra Kupsh died of diphtheria at a Christian Scientist camp in Colorado. In 1985, three students died of measles at Principia College, a Christian Scientist school in Elsah, Missouri. At the time, the CDC's Walter Orenstein noted that the high rate of measles deaths at Principia resembled "the kind of mortality statistics we see in the Third World." The 1985 outbreak wasn't the end of it; four more measles outbreaks occurred among Christian Scientist students in the St. Louis area between 1985 and 1994.

But the incident that attracted the most attention occurred in the fall of 1972, when a Christian Scientist high school in Greenwich, Connecticut, suffered a polio outbreak. Eleven children were paralyzed. At the time of the outbreak, not a single case of polio had occurred in Connecticut in more than three years. One health commissioner was so shaken that he wrote an editorial in the *New England Journal of Medicine*: "Although I am not arguing for absolute state control of the lives of private citizens, I am deeply bothered by the fact that disease prevention measures of documented benefit can be withheld from children by their parents in the name of religious freedom, jeopardizing not only the health and lives of the children so denied but those of the community as well. The courts of this land have long since set precedent in the protection of children from the irresponsible acts of their parents."

The successful lobbying of Christian Scientists to include a religious exemption in New York State changed the strategy of those who wanted to avoid vaccines. Two seminal rulings opened the floodgates. The first involved William Maier and his three children—all of whom had been excluded from Fabius Central School in Onondaga County, New York. Maier argued that, although he wasn't a Christian Scientist, he, like they, believed that "the sanctity of the human body cannot be violated by injection," and that his beliefs were entitled to the same protection as a Christian Scientist's. The judge agreed. The second ruling allowing religious exemptions involved Beulah Dalli, a mother in Lowell, Massachusetts. Dalli argued that her six-year-old daughter, Belinda, had been unfairly excluded from school because she wasn't vaccinated. Although Dalli, like Maier, wasn't a Christian Scientist, she believed that the use of vaccines was a violation of the Bible and its teachings, specifically the admonition to "keep the body clean and acceptable to God." The court ruled in her favor, stating that it was unfair to allow vaccine exemptions for some religious groups but not others. By 2009, forty-eight states allowed religious exemptions to vaccination.

Religious exemptions opened the door to philosophical exemptions. A critical ruling in the late 1980s involved two Northport, New York, couples: the Sherrs and the Levys. Alan Sherr, a chiropractor, argued that his religion—a mail-order church in Sarasota, Florida, called the Missionary Temple at Large of the Universal Religious Brotherhood, Inc.—believed that vaccines represented an unwanted intrusion into the body. However, because Sherr had circumcised his son and allowed dentists to fill his son's cavities, the intrusion argument didn't hold. The Levys, on the other hand, were more compelling. Louis Levy argued that the state should not be allowed to force his daughter, Sandra, to be vaccinated because "We feel that any introduction into the process of a foreign element outside the normal processes of the body is going to affect the body adversely and, therefore, we feel it is a violation in a sense of our nature—physical, spiritual, religious nature." The judge ruled in Levy's favor, stating that vaccine exemptions could be granted if "beliefs [were] held with the strength of religious convictions," even if parents weren't members of a religious group. By 2010, twenty-one states allowed philosophical exemptions to vaccination.

By the late 1890s, the anti-vaccine movement in England had spawned the Anti-Compulsory Vaccination Act. If citizens didn't want to be vaccinated, they could obtain a certificate of conscientious objection. As more and more citizens opted out, England became the United Kingdom's epicenter of smallpox disease. By the early twenty-first century, American anti-vaccine activists—through passages of religious and philosophical exemptions—had also impacted the public's health. Between 1999 and 2009, four studies examined whether the freedom to avoid vaccines also meant the freedom to catch and transmit infections.

In 1999, Daniel Salmon and co-workers from the Johns Hopkins School of Public Health found that the risk of contracting measles in five- to nine-year-olds whose parents had chosen not to vaccinate them was one hundred and seventy times greater than for vaccinated children.

In 2000, Daniel Feikin and his colleagues from the Respiratory Diseases Branch at the CDC found that three- to ten-year-olds in Colorado whose parents had chosen not to vaccinate them were sixty-two times more likely to catch measles and sixteen times more likely to catch whooping cough. They also found that measles outbreaks were more common in schools with greater numbers of unvaccinated children.

In 2006, Saad Omer and his colleagues, also from the Johns Hopkins School of Public Health, examined the impact of philosophical exemptions. Between 1991 and 2004, the number of unvaccinated children in states with philosophical exemptions more than doubled. Children in states with easy-to-obtain exemptions (granted by the simple signing of a form) were almost twice as likely to suffer outbreaks of whooping cough.

In 2009, Jason Glanz and his co-workers from the Institute for Health Research in Denver—confirming previous studies—found that unvaccinated children in Colorado were twenty-three times more likely to suffer whooping cough.

The impact of anti-vaccine activism hasn't been limited to England and the United States. Studies have shown that the risk of whooping cough was ten to a hundred times greater in countries where immunization had been disrupted by anti-vaccine movements (such as Sweden, Japan, the Russian Federation, Ireland, Italy, and Australia) than in countries that had maintained high vaccination rates (such as Hungary and Poland).

The results were in. A choice not to get a vaccine was a choice to increase the risk of suffering and possibly dying from an infectious disease. Worse still, it was a choice to put one's neighbor at greater risk. So why are more and more parents choosing not to vaccinate their children? The answer can be found in part in the writings of a professor of biology at the University of California at Santa Barbara who has explained why, under certain circumstances, a choice not to get a vaccine is far more rational than a choice to get one.

In 1968, Garrett Hardin published an essay in the journal *Science* called "The Tragedy of the Commons." Hardin was interested in the problem of population control, but his observations can easily be applied to the problem of vaccine refusal. "Picture a pasture open to all," he wrote. "It is to be expected that each herdsman will try to keep as many cattle as possible on the commons. As a rational being, each herdsman seeks to maximize his gain. He asks, 'What is the utility *to me* of adding one more animal to my herd?' The utility has one negative and one positive component."

Hardin continued: "The positive component is a function of the increment of one animal. Since the herdsman receives all the proceeds from the sale of the additional animal, the positive utility is nearly +1. The negative component is a function of the additional overgrazing created by one more animal. Since, however, the effects of overgrazing are shared by all the herdsmen, the negative utility for any particular decision-making herdsman is only a fraction of −1."

Imagine a herdsman as a parent choosing not to vaccinate a child. The great unsaid about vaccines is that if everyone in the world is vaccinated, it would make more sense for a parent not to vaccinate. This is true for two reasons. First: as more and more children are vaccinated it becomes less and less easy for viruses and bacteria to spread. Indeed, when enough people are vaccinated, these infections simply stop spreading. For example, when the polio vaccine was introduced in the United States in 1955, only 40 percent of the population was immunized, and, although the number of people paralyzed by polio declined, the virus continued to spread. However, after 70 percent of the population had been immunized, the virus stopped spreading and polio infections were eliminated from the United States. The same is true for measles. The only difference is that measles is much more contagious than polio, so a larger percentage of people (about 95 percent) needed to be immunized. After enough people are immunized, those who aren't can hide in the herd, protected by those around them. Second: although vaccines are safe, they aren't perfectly safe. All vaccines have side effects, mostly pain and tenderness at the site of injection; but some side effects, such as allergic reactions, can be quite severe. By choosing not to vaccinate, one can enjoy the benefits of hiding in the herd without risking such rare but real side effects. (This is true for all vaccine-preventable diseases save one: tetanus, which is acquired from the soil, not from another person. So, even if everyone in the world is vaccinated against tetanus except for one child, the risk to that child of acquiring tetanus is the same.)

Hardin continued his essay by explaining how a rational choice can become an irrational one: "The rational herdsman concludes that the only sensible course for him to pursue is to add another animal to the herd. And another; and another. . . . But this is the conclusion reached by each and every rational herdsman sharing the commons. Therein is the tragedy. Each man is locked into a system that compels him to increase the herd without limit—in a world that is limited. Ruin is the destination toward which all men rush, each pursuing his own best interest in a society that believes in the freedom of the commons. Freedom in a commons brings ruin to all."

Such is the case with vaccines. As more and more people have chosen not to vaccinate, herd immunity has broken down. Now, a choice not to get a vaccine has the benefit of avoiding rare side effects, but not the benefit of herd immunity. The studies of Salmon,

Feikin, Omer, and Glanz showed that the choice not to get measles or pertussis vaccine was a choice to risk infection because not enough people were getting vaccinated.

No one has been hit harder by the loss of an immunological commons than children who can't be vaccinated.

On October 20, 2009, Stephanie Tatel, an elementary school reading specialist in Charlottesville, Virginia, published an article on Slate.com on her efforts to find a child-care center for her son. "Last year, while searching for child care for our 2-and-a-half-year-old son, my husband and I thought we had found the perfect arrangement," she wrote, "an experienced home day care provider whose house was an inviting den of toddler industriousness. Under her magical hand, children drifted calmly and happily from the bubble station to the fairy garden to the bunnies and the trucks, an orchestrated preschool utopia. But when I asked, 'Are any of the children here unvaccinated?' the hope of my son's perfect day care experience burnt to a little crisp. As it turned out, one child had a philosophical or religious exemption—a convenient cover-all exemption that many doctors grant, no questions asked, when a parent requests one. I still do not understand how the state can allow one to attribute his or her fear of vaccines and their unproven dangers to religion or philosophy. Ordinarily I wouldn't question others' parenting choices. But the problem is literally one of live or don't live. While that parent chose not to vaccinate her child for what she likely considers well-founded reasons, she is putting other children at risk. In this instance, the child at risk was my son. He has leukemia."

Tatel knew that the unvaccinated child posed a risk to her son. "I realize that anti-vaccine sentiment has been around as long as the vaccines themselves," she wrote. "But I wonder whether they have fully considered that the herd immunity, of which they are taking advantage, is designed to protect those who cannot be vaccinated. For now, we will hire an at-home sitter for [our son]. When he is ready to go off to school, we will have to face this issue again. Because we want him to have as 'normal' a life as possible, we'll likely send him off in the bright yellow school bus and cross our fingers that the kid sitting next to him didn't just attend a 'chickenpox party' over the weekend. Because what's 'just a case of chickenpox' for that kid could be a matter of life or death for mine."

Like Stephanie Tatel's son, hundreds of thousands of people in the United States cannot be vaccinated, forced to depend on those around them to be protected.

In 1998, Hardin wrote another essay titled "Extension of 'The Tragedy of the Commons.'" In the intervening thirty years, Hardin had witnessed increasing pollution of the air, seas, and land by "herdsmen" who had continued to "overgraze." His summary of the situation was poignant. "Individualism is cherished because it produces freedom," he wrote. "But the gift is conditional."

CHAPTER 9

The Mean Season

You know, you remove certain medications off
shelves because they're deemed unsafe. Why not
vaccines?

—JENNY MCCARTHY ON
LARRY KING LIVE, APRIL 3, 2009

The breakdown in herd immunity in the United States at the beginning of the twenty-first century hasn't silenced anti-vaccine
activists. Although Barbara Loe Fisher and her National Vaccine
Information Center have been heard from less frequently, other
groups have stepped in to take her place—specifically the Coalition
for Vaccine Safety. Formed from a variety of groups that believe
vaccines cause autism, this new breed of anti-vaccine activism has
a dramatically different style: meaner, cruder, more strident, and
less professional.

Jenny McCarthy was born on November 1, 1972, in Chicago, Illinois. She attended St. Turibius Grade School on Chicago's south side
and Mother McAuley Liberal Arts High School. Later, she entered
Southern Illinois University in Carbondale to study nursing. But her
heart wasn't in it; she wanted to be a model.

Her success was immediate. In October 1993, McCarthy was
Playboy magazine's Playmate of the Month; in 1994, she was Play-
mate of the Year. Her affiliation with *Playboy* didn't end there. Mc-
Carthy hosted the *Playboy* television show *Hot Rocks*, which

featured uncensored music videos, and the dating show *Singled Out*. In 1996, she landed a bit part in the comedy *The Stupids*. That same year, *People* magazine named her one of the fifty most beautiful people in the world.

McCarthy's movie career wasn't limited to *The Stupids*. In 1998, she had a small role in *BASEketball* and the following year in *Diamonds*, directed by John Asher, whom she married in September 1999. A few years later, on May 18, 2002, their only child, Evan, was born in Los Angeles. But all was not well. Following a chance encounter with a stranger, McCarthy knew that something was different about her son. "One night I reached over and grabbed my Archangel Oracle tarot cards and shuffled them and pulled out a card," she wrote. "It was the same card I had picked over and over again the past few months. It was starting to drive me crazy. It said that I was to help teach the Indigo and Crystal children. [Later,] a woman approached Evan and me on the street and said, 'Your son is a Crystal child,' and then walked away. I remember thinking, 'Okay, crazy lady,' and then I stopped in my tracks. Holy shit, she just said 'Crystal child,' like on the tarot card." McCarthy realized that she was an Indigo adult and Evan a Crystal child. Although Evan would soon be diagnosed with autism, McCarthy took heart in the fact that Crystal children were often mislabeled as autistic. According to Doreen Virtue, author of *The Care and Feeding of Indigo Children*, "Crystal Children don't warrant a label of autism! They aren't autistic, they're AWE-tistic."

In 2005, McCarthy changed her mind. She abandoned her tarot-card predictions and embraced the notion that her son was autistic and that vaccines were responsible. On September 18, 2007, in front of millions of viewers on *Oprah*, she described the moment that changed her life: "Right before my son got the MMR shot I said to the doctor, 'I have a very bad feeling about this shot. This is the autism shot, isn't it?' And he said, 'No! That is ridiculous. It is a mother's desperate attempt to blame something on autism.' And he swore at me. And then the nurse gave him that shot. And I remember going, 'Oh, God, no!' And soon thereafter I noticed a change. The soul was gone from his eyes." By 2007, researchers had pub-

Jenny McCarthy with then husband John Asher at a party at the Playboy mansion. McCarthy has become America's most-recognized anti-vaccine activist. (Courtesy of Kenneth Johansson/Corbis.)

lished several studies showing that MMR didn't cause autism; McCarthy was unconvinced. "My science is Evan, and he's at home," she said. "That's my science."

Using the fame of her *Playboy* and movie career, McCarthy soon became America's most recognized anti-vaccine crusader.

In many ways, Jenny McCarthy and Barbara Loe Fisher are similar. Both dramatize their personal stories in vivid, heart-wrenching terms. Where Fisher describes her son's learning disabilities as brain damage, McCarthy likens certain symptoms of her son's autism to death. When asked during an interview on CNN whether her campaign against vaccines could result in children dying from preventable infections, McCarthy said, "People are also dying from vaccination. Evan, my Evan, my son died in front of me for two minutes." McCarthy also shares Fisher's disdain for public health officials and pharmaceutical companies. "I think they need to wake

up and stop hurting our kids," she said. Finally, both Fisher and Mc-
Carthy continually reshape their messages to fit the style of the time.
Fisher switched from a campaign against pertussis vaccine to one
against all vaccines, claiming they caused chronic diseases. McCarthy,
supported by environmental activists like Robert F. Kennedy, Jr., and
Don and Deirdre Imus, later decided that her son's autism wasn't
caused by MMR. It was caused by vaccine toxins—specifically, mercury, aluminum, and anti-freeze. (McCarthy later undercut her stop-
injecting-toxins-into-our-bodies message by saying, "I love Botox. I
absolutely love it. I get it minimally, so I can still move my face. But
I really do think it's a savior." Made by the bacterium that causes
botulism, *bot*ulinum *tox*in [Botox] is one of the world's most powerful toxins.)

Although Jenny McCarthy and Barbara Loe Fisher are alike in several ways, their differences are striking.

Unlike Fisher, McCarthy often resorts to profanity. On April 1,
2009, Jeffrey Kluger, a veteran scientific correspondent, interviewed
McCarthy for *Time* magazine. Kluger, who had recently written a
popular book about the polio vaccine, asked, "What about the polio
clusters in unvaccinated communities like the Amish in the United
States? What about the 2004 outbreak that swept across Africa and
Southeast Asia after a single province in northern Nigeria banned
vaccines?" McCarthy replied, "I do believe sadly it's going to take
some diseases coming back to realize that we need to change and
develop vaccines that are safe. If the vaccine companies are not lis-
tening to us, it's their fucking fault that the diseases are coming
back. They're making a product that's shit. If you give us a safe vaccine, we'll use it. It shouldn't be polio versus autism." Kluger also
asked McCarthy about the measles vaccine. "And yet in many cases,
vaccines have effectively eliminated diseases," he said. "Measles is
among the top five killers in the world of children under five years
old, yet it kills virtually no one in the United States thanks to vaccines." McCarthy replied, "If you ask a parent of an autistic child
if they want the measles or the autism, we will stand in line for the
fucking measles."

Unlike Fisher, McCarthy is comfortable dispensing medical advice. In a fifteen-minute video designed for parents, she explains what causes autism and how to treat it: "Autism is a toxic overload. And one of the things I want you to write down and then put on your refrigerator are just five things: food, supplementation, detox, medicines, and positive thinking." McCarthy starts with food, explaining that children should avoid gluten (wheat, barley, or rye) and casein (dairy products). "When you can't break it down," she says, "they get stoned, which accounts for their moods, their spaciness, their addiction for things. The mom says he just has to have his twelve cups of milk a day; he just has to have his mac and cheese. Well, no kidding. You know, I really liked my marijuana in college, too. When they want that milk and they want that wheat, you're giving them a joint."

Although both McCarthy and Fisher openly despise pharmaceutical companies, McCarthy promotes their products. On her video describing how to treat autism, McCarthy says, "Some of our kids can't absorb the nutrients that we give them so they have to be supplemented. Some of the multi-vitamins that I like are Super Nu-Thera® that can be found at Kirkman Laboratories. [McCarthy displays a picture of Super Nu-Thera® followed by Kirkman's Web site.] Culturelle® you can find at any pharmacy; it's over-the-counter. ThreeLac® is one of my favorites because it's a probiotic that also eats yeast and pretty much recovered Evan. Also found at Kirkman. If you're unsure about dosage, ask your pediatrician; but most of the time they don't know anything. So I would say ask someone at Kirkman." At the end of the video, McCarthy promotes another pharmaceutical company with the statement, "For more information about vitamins visit www.kartnerhealth.com." According to McCarthy's logic, then, those who promote vaccines are evil because they're fronting for products that gross $17 billion a year; while those who promote supplements are virtuous because they're fronting for products—almost all of which are of unproved efficacy—that gross $80 billion a year.

Perhaps the most important difference between Jenny McCarthy and Barbara Loe Fisher is their backers. Both are heartily endorsed

by personal-injury lawyers with much to gain from the Vaccine Injury Compensation Program; but McCarthy, not Fisher, is supported
by a wealthy financier. On April 3, 2009, she appeared on *Larry
King Live*. At the end of her segment, King asked, "Jenny, you have
a Web site. What is it?" Although McCarthy has her own Web site,
she didn't mention it. "Generationrescue.org," she said. "You can
go there for more information."

Generation Rescue was started by a venture capitalist named
J. B. Handley, who believes his son's autism was caused by thimerosal
in vaccines. Like McCarthy, Handley has a cure: chelation, a potentially dangerous therapy of unproved efficacy that helps rid the
body of heavy metals like mercury and lead. (In 2005, a five-year-
old autistic boy died during chelation in suburban Pittsburgh.) To
proselytize the miracle of chelation, Handley recruited a group of
parents to spread the word, calling them Rescue Angels. Generation
Rescue's mission is, in part, to "gather the information that currently exists about mercury toxicity and publicize the truth so parents can make the best decision to help their children heal." The key
word in Generation Rescue's mission statement is *publicize*. On June
8, 2005, Handley's organization took out a full-page ad in the *New
York Times*. At the top of the page, in bold, black type, the ad declared, "MERCURY POISONING AND AUTISM: IT ISN'T JUST
A COINCIDENCE." On April 6, 2006, Handley's organization
took out another full-page ad, this time in *USA Today*. Written in
letters two inches high, the ad angrily stated, "IF YOU CAUSED A
6,000% INCREASE IN AUTISM WOULDN'T YOU TRY TO
COVER IT UP, TOO? IT'S TIME FOR THE CDC TO COME
CLEAN WITH THE AMERICAN PUBLIC." On February 25,
2009, Generation Rescue took out yet another full-page ad in *USA
Today*. This time Handley wanted to alert the public about Bailey
Banks, a boy whose parents had claimed that vaccines had caused
his autism. "A LITTLE BOY SHOULDN'T HAVE TO TAKE ON
AN ENTIRE INDUSTRY ALONE. IT'S TIME THE GOVERN-
MENT TOLD THE TRUTH ABOUT CHILDHOOD VAC-
CINES." Each of these ads costs as much as $180,000. Generation
Rescue is an advertising arm of the anti-vaccine movement.

Apparently McCarthy's stance against vaccines impressed Handley. So, in 2009, "Generation Rescue" became "Jenny McCarthy's Autism Organization—Generation Rescue," complete with pictures of and messages from McCarthy.

Handley brought something to the anti-vaccine movement that hadn't been seen before: personal intimidation. He didn't just rail against journalists or professional societies or vaccine advocates; he sued them or sent them hate-filled emails or maintained Web sites to vilify them or screamed at them on national television. On CBS's daytime program *The Doctors*, Handley, appearing with McCarthy, attacked the show's host, Dr. Travis Stork. Stork was convinced by studies that had exonerated vaccines as a cause of autism; Handley wasn't:

> STORK: In my opinion—and this is just me wanting to have an open debate about this—vaccines are really the one thing we *have* looked at as causing autism.
> HANDLEY: That is completely bogus! That is such a bogus statement!
> STORK: No, that's . . .
> HANDLEY: How many vaccines have they looked at in these studies?! How many?! What's the answer?! I'm so sick of doctors who don't read the studies, who don't know the details sitting here telling parents and reassuring them that vaccines don't cause autism. It's irresponsible.

Stork was angry that Handley had chosen to characterize doctors as uncaring and falsely reassuring.

> STORK: And this is the biggest problem, and the reason that doctors in this country are frustrated.
> HANDLEY: Read the science!
> STORK: All you're doing is you're antagonizing a medical community that wants to help these kids. OK?
> HANDLEY: You haven't done the research.
> STORK: You're antagonizing me. Why would you do that?
> HANDLEY: Because my son was . . .

J. B. Handley takes on Dr. Travis Stork on an episode of *The Doctors*, May 6, 2009. (Courtesy of *The Doctors*, Stage 29 Productions, and CBS Television Distributions.)

STORK: OK! Everyone wants to blame someone, right? What we're trying to figure out here is how to help kids. But all you do when you yell at me on my stage, all you do is anger me.

HANDLEY: I'm sorry I hurt your feelings, but you don't know the details.

STORK: I asked you to defend your stance, and all you did was attack me as an individual. Why would I want to listen to you when you do that to me?

On an anti-vaccine Web site, Handley boasted about his ability to take on doctors. "I'm not intimidated by any of these jokers," he wrote. "Their degrees mean zippo to me, because I knew plenty of knuckleheads in college who went on to be doctors, and they're still knuckleheads."

Later in *The Doctors* program Stork revealed how, on the strength of McCarthy's star power, she had rigged the show.

MCCARTHY: Go call the AAP [American Academy of Pediatrics] and see if they'll sit down with us and they'll say, 'No, tell them to write a letter.'

STORK: Let me just say this openly to everyone. You know, we wanted to have someone from the AAP here today, but you refused to allow them to come. So if you want to engage them in a debate, they would have been here.

Following the show, David Tayloe, a North Carolina pediatrician and president of the AAP, wrote to Lisa Williams, producer of *The Doctors*.

Dear Ms. Williams,

Once again, Jenny McCarthy has struck a blow to public health, and "The Doctors" have given her the loudspeaker. I was disappointed with the May 6 episode featuring Ms. Mc-Carthy and her associates from the anti-vaccine group Generation Rescue.

True . . . Dr. Travis Stork revealed Ms. McCarthy's hypocrisy over her unfounded claims that the American Academy of Pediatrics has refused to sit down with her. In fact, she is the one who refused to engage the AAP in an honest dialogue on the show. That only begs the question: Why allow her this platform at all? I do not understand why you granted a celebrity the power to veto a guest who is actually prepared to refute her unscientific claims. The casual viewer would not have known that the deck was stacked; every guest, including J. B. Handley . . . plays a prominent role in Generation Rescue. This is a small, vocal minority. But a young parent watching your show would get the mistaken impression they represent the consensus on vaccines. This misinformation is costly. Unimmunized children are dying of vaccine preventable diseases in this country. . . .

I fear you let drama and ratings trump sound medical advice.
Sincerely,
David T. Tayloe, Jr., M.D.
President, American Academy of Pediatrics

Handley appears to embrace the legal aphorism "If the law is on your side, argue the law; if the facts are on your side, argue the facts; if neither is on your side, attack the witness." The scientific evidence against him, Handley chose ad hominem attacks, arguably the lowest form of debate. His confrontation with Travis Stork was one of many:

- On January 31, 2008, Nancy Minshew, a professor of psychiatry and neurology at the University of Pittsburgh and director of a Center of Excellence in Autism of the National Institutes of Health (NIH), confronted anti-vaccine activists. In an article in the *Pittsburgh Post-Gazette* titled "Pitt Expert Goes Public to Counter Fallacy on Autism," Minshew said, "The weight of the evidence is so great that I don't think there is any room for dispute. I think the issue is done." Handley, who had become aware of an email exchange between Minshew and a parent of a child with autism, threatened to post it on an anti-vaccine Web site. Minshew was upset, writing, "Mr. Handley, none of you have permission to share emails that I have sent to you as individuals. Unlike the newspaper, which was public, private emails to individuals sent confidentially are not for public quotation." Handley responded immediately. "Says who?" he wrote. "And tough shit."
- In an interview with *Cookie* magazine in their August 2008 issue, actress Amanda Peet, who starred in *2012*, *A Lot Like Love*, *Martian Child*, *Something's Gotta Give*, *Syriana*, *The Whole Ten Yards*, and *X-Files*, talked about the importance of vaccines. Peet was worried about the growing number of unvaccinated children in southern California; worried about how it might affect her young daughter. "I was shocked at the amount of misinformation [about vaccines] floating around," she said, "particularly in Hollywood." Again, Handley resorted to threats: "Ms. Peet, I have a quick message for you: you have no idea who you are messing with."
- On August 4, 2008, Handley attacked Every Child by Two (ECBT), a nonprofit organization founded by Betty Bumpers

and Rosalynn Carter. Questioning their funding sources, Handley wrote, "By non-profit standards, ECBT is a rat-shit organization."

- On September 10, 2008, after yet another study had found no evidence that MMR caused autism, Geri Dawson, chief science officer for the advocacy organization Autism Speaks, sent out a press release reassuring parents about the safety of the vaccine. Handley's response was personal: "Geri Dawson is either a blithering idiot or she is a corrupt partisan hack who so desperately wants the autism-vaccine thing to just die so she can get back to work chasing her genetic-psychological theories on autism that she will happily go along with the mainstream spin on a stupid little study."

- On October 30, 2008, after NBC's Dr. Nancy Snyderman appeared on *The Today Show* supporting the science showing no link between vaccines and autism, Handley called her "NBC's pharma-whore in residence."

- On December 15, 2008, in a blog entry titled "Some *New York Times* Reporters Are Just Ignorant," Handley attacked Gardiner Harris, who had written an article exonerating vaccines as a cause of autism. Handley wrote, "There's a reporter named Gardiner Harris who writes for the *New York Times*. I've probably talked to a hundred or so reporters in my time and he is unquestionably the biggest jackass I have ever encountered."

- In November 2009, a freelance reporter named Amy Wallace wrote an article for *Wired* magazine titled "An Epidemic of Fear." The cover of *Wired* featured an infant staring out from behind the word *FEAR* in three-inch high letters; the subheading read, "Vaccines don't cause autism. But some panicked parents are skipping their baby shots. Why that bad decision endangers us all." As he had done with Nancy Snyderman, Handley demeaned Wallace, a single mother living in southern California, with a sexual reference, implying that she had been intellectually raped using a date-rape drug. After Wallace described his comments on National Public Radio, Handley called her a "cry baby." Wallace lamented "the way people like Handley use gender and sexuality as weapons to bully their opponents," arguing that "the debate needs to be civil. That's part of what I've been trying to participate in—a civil discussion of these issues."

- On January 12, 2010, Handley went after those he believed were most responsible for causing autism—pediatricians: "If a doctor sticks six vaccines into a child while the child is taking antibiotics for an ear infection and Tylenol for a cold, he's not a doctor, he's a criminal, and should be hauled into jail on the spot for assault and battery. If the child also happens to have eczema, long-term diarrhea, and has missed a milestone or two, perhaps the charges should be attempted murder."

Handley's appearance on entertainment television didn't end with *The Doctors*. On April 3, 2009, Handley, appearing on *Larry King Live*, said, "They [the AAP] rubber stamp every vaccine on the schedule. Dr. [Margaret] Fisher [representing the AAP on the show] never answered why so few companies have picked up varicella, flu, rotavirus. Meantime, AAP rubber stamps every vaccine, like Gardasil [human papillomavirus vaccine], which is damaging teenaged girls right now; which will likely be pulled from the market very soon." Handley had mischaracterized the American Academy of Pediatrics, failing to account for the enormous amount of work done by the AAP's Committee on Infectious Diseases before recommending vaccines. Further, he implied that few companies manufactured vaccines against varicella (chickenpox), influenza, or rotavirus because they weren't safe or effective. But that's not the reason; only a few companies make vaccines because vaccines, compared with drugs, are enormously expensive to test and manufacture. Patricia Danzon, an economics professor at the Wharton Business School at the University of Pennsylvania, has publicly expressed surprise that more than one company makes *any* vaccine— given that they're used only once or a few times in a lifetime. But Handley's most outrageous comment was that Gardasil was not only dangerous (a contention refuted by careful study) but also about to be pulled from the market. This was clearly untrue. Given that Gardasil prevents the only known cause of cervical cancer—and given that at least some CNN viewers, believing Gardasil was about to be withdrawn, might have chosen not to give the vaccine to their daughters—Handley's statement tested the bounds of free speech. Citizens of the United States are not allowed to shout "Fire!" in a crowded movie theater because it puts lives at risk. Arguably, Handley's baldly inaccurate statement about Gardasil's imminent withdrawal might have caused some parents to withhold the vaccine for their daughters, putting them at needless risk of cervical cancer.

Handley's disregard for established science didn't end with his comments about Gardasil. On the May 6, 2009, episode of *The Doctors*, he said, "As mad as we are—and we are mad and frustrated that we aren't heard—we want to find common ground. We don't want all these deadly childhood diseases to return, either. One of the recommendations that we've made to parents is go back to the 1989 schedule, before it became—in our opinion—overly commercialized." Handley was asking to go back to a time when, every year, pneumococcus caused tens of thousands of cases of pneumonia and thousands of cases of meningitis, killing about two hundred children; when hepatitis B virus infected about sixteen thousand young children; when rotavirus caused seventy thousand babies to become so dehydrated that they had to be hospitalized; and when Hib caused twenty thousand children to suffer bloodstream infections, pneumonia, epiglottitis, or meningitis. Handley's advice to return to a decade when hundreds of thousands of children were harmed by what are now preventable infections ranks as one of the most irresponsible, ill-informed statements ever made by an anti-vaccine activist.

Handley's notion of a gentler, better time before vaccines wasn't new. One month earlier he had made a similar suggestion. During his April 2009 appearance on *Larry King Live*, Handley, referring to the notion that babies were getting too many vaccines too soon, said, "Larry, we have no idea what the combination risk of our vaccine schedule looks like. At the two-month visit, a child gets six vaccines in under fifteen minutes. The only way to test that properly would be to have a group of kids who get all six and a group of kids who got none and see what happens. They don't do that testing. They have no idea." Handley was asking for a study of vaccinated and unvaccinated children. One result is certain: given recent outbreaks of Hib, measles, mumps, and pertussis, unvaccinated children would suffer and possibly die from preventable infections. It would be, of course, an entirely unethical experiment. No investigator could prospectively study children who are denied a potentially lifesaving medical product. And no university's or hospital's institutional review board worth its salt would ever approve such a study. Handley's proposal harkens back to a dark time in our history when—between 1932 and 1972—investigators prospectively studied four hundred African-American men from one of the poorest counties in Alabama to see what would happen if their syphilis went untreated. It was called the Tuskegee Study. Withholding antibiotics that could have cured them, it was probably the most unethical medical experiment ever performed in America.

Another obvious difference between Jenny McCarthy and Barbara Loe Fisher is that McCarthy is a celebrity. It's her celebrity that has landed her on shows like *Oprah* and *Larry King Live*. And it's her celebrity that has enabled her to determine the guest list. McCarthy isn't the first person to use fame to influence the public about vaccines. In the 1950s, the March of Dimes promoted polio vaccines using singers Elvis Presley, Bing Crosby, Judy Garland, and Frank Sinatra, comedians Jack Benny and Lucille Ball, ventriloquist Edgar Bergen and his puppet Charlie McCarthy, and actors Clayton Moore (the Lone Ranger), Mary Pickford (America's sweetheart), and Mickey Rooney; even Mickey Mouse participated ("Hi ho, hi ho, we'll lick that polio"). The tradition is still alive. In addition to Amanda Peet, actors such as Keri Russell (*Waitress*, *August Rush*) and Jennifer Garner (*Juno*, *13 Going on 30*) have spoken on behalf of vaccines. So has Heisman trophy winner Archie Griffin. But unlike in the 1950s, many celebrities today use their fame to scare the

Jim Carrey exhorts crowd at an anti-vaccine rally in front of the Capitol, June 4, 2008. (Courtesy of Getty Images.)

public. In addition to McCarthy, Jessica Alba, Cindy Crawford, Matthew McConaughey, Doug Flutie, and Aidan Quinn have all said that vaccines are unsafe. But no actor has joined the fray more than Jim Carrey. Unlike McCarthy, Carrey is recognized by most Americans for his work in the popular comedies *Liar, Liar*; *Ace Ventura: Pet Detective*; and *Dumb and Dumber* as well as for his serious roles in *The Truman Show* and *The Majestic*. People know Jim Carrey as a warm and funny man; they trust him. So when Carrey started to date Jenny McCarthy, and to share her anti-vaccine passion, her star power and the impact of her message increased dramatically.

McCarthy and Carrey were a compatible anti-vaccine couple, sharing the notion that vaccines are a conspiracy run by pharmaceutical companies. On April 3, 2009, on *Larry King Live*, Carrey said, "The AAP is financed by drug companies. Medical schools are financed by drug companies." Carrey also doesn't trust public health agencies. "I don't think people that are charged with the public health any longer have our best interests at heart all the time," he said. "Parents have to make their own decisions: educated decisions." Unfortunately, like McCarthy and J. B. Handley, Carrey does little to educate them. On the same segment of *Larry King Live*, Handley noted that "twenty-seven countries chose not to vaccinate for the chickenpox." Carrey knew why. "That vaccine doesn't work," he said.

Handley's implication that chickenpox is unimportant and Carrey's statement that the vaccine doesn't work are inconsistent with the evidence. The chickenpox vaccine was first licensed and used in the United States in 1995. Although many people think of chickenpox as a benign disease—a simple rite of childhood passage—it isn't; every year chickenpox causes children to be hospitalized and to die. The virus, which disrupts the skin with painful blisters, allows entrance of bacteria like *Streptococcus pyogenes*. Dubbed "flesh-eating bacteria" by the press, streptococcus causes serious and occasionally fatal diseases like necrotizing fasciitis (a deep-seated infection that dissects rapidly through muscles, necessitating emergency surgery) and pyomyositis (in which muscles liquefy from massive inflammation). The virus can also travel to the lungs causing pneumonia and to the brain causing encephalitis. Worst of all: you never get rid of chickenpox. Even after people recover from the infection, the virus lives silently in nerve roots, occasionally reawakening later in life causing shingles, one of medicine's most debilitating diseases. Shingles is so painful that it has at times led to suicide. And shingles doesn't only affect the skin; sometimes when the virus reawakens it causes strokes, resulting in permanent paralysis. Chickenpox is a disease worth preventing. And, thanks to the chickenpox vaccine, American children are now much less likely to catch it. Since the vaccine was released, twenty studies—performed between 1997 and 2006—have evaluated whether the vaccine works. Every one of them found that it did. Not surprisingly, the number of children with chickenpox—once totaling about four million a year—has declined dramatically. But Jim Carrey never mentioned these data. Rather, he declared to several million people on national television that the vaccine didn't work—a statement that was entirely false and went completely unchallenged.

In October 2009, during the swine flu (H1N1) epidemic, another celebrity threw his hat into the ring: Bill Maher, the popular host of HBO's *Real Time with Bill Maher*. For those who followed him on Twitter, Maher advised that getting a vaccine to prevent swine flu was for "idiots." On his television show, he debated Bill Frist, a heart surgeon and former Senate majority leader from Tennessee.

> MAHER: Why would you let them [doctors] be the ones to stick a disease into your arm? I would never get a swine flu vaccine or any vaccine. I don't trust the government, especially with my health.
> FRIST: On the swine flu, I know you really believe that. And let me just . . .
> MAHER: You say that like I'm a crazy person.

Frist told the story of a healthy thirty-year-old man who had died of swine flu in his (Frist's) hospital. Maher didn't buy it.

> MAHER: This is not a serious flu. Let's be honest. There must be something more to this. I cannot believe that a perfectly healthy person died of this swine flu. That person was not perfectly healthy. Western medicine misses a lot.

Frist told Maher about two recent publications in the *New England Journal of Medicine* describing the high risk of fatal influenza in pregnant women.

> FRIST: I know you don't believe this, but I'm telling you the facts. Because if you send a signal out telling pregnant women not to get this vaccine . . .
> MAHER: I do.
> FRIST: Well, you're wrong. I'm serious.

One month later, Maher wrote an article for the *Huffington Post* titled "Vaccination: A Conversation Worth Having." In it, he sounded many classic anti-vaccine themes: for example, that vaccines contain

Bill Maher, host of
HBO's *Real Time with Bill
Maher*, declared that the
novel H1N1 (swine flu)
vaccine was for "idiots."
(Courtesy of Kabik/
Retna Ltd./Corbis.)

dangerous additives ("the formaldehyde, the insect repellent, the
mercury"), that diseases prevented by vaccines were disappearing
anyway ("polio had diminished by over 50 percent in the thirty
years before the vaccine"), and that common belief is common wisdom ("sixty-five percent of the French people don't want [the flu
vaccine]. Are they all crazy too?"). Then Maher gave his readers the
source of his information: "Someone who speaks eloquently about
this is Barbara Loe Fisher, founder of the National Vaccine Information Center. I find her extremely credible, as I do Dr. Russell
Blaylock, Dr. Jay Gordon, and many others. But I shouldn't have
even mentioned them because I don't want to be 'The Vaccine
Guy'!! Look it up yourself and stop asking me about it. I'm already
'The Religion Guy,' and that's enough work!"

When Maher called himself "The Religion Guy," he was referring to his 2008 movie, *Religulous* (presumably a contraction of the
words *religion* and *ridiculous*). Maher took on religion, claiming
that religious beliefs weren't supported by scientific evidence. At the
beginning of the movie he asked, "Why is believing something without evidence good?" Maher noted that many scientists were either

atheists or agnostics. He was likening himself to them. But that's where the similarity ended.

Maher argued that influenza vaccine was equivalent to "sticking a disease into your arm." Following his criticism, Maher received letters from doctors explaining how the influenza vaccine was made and how it works—and why it wasn't like "sticking a disease into your arm." But Maher didn't need their help. "I read *Microbe Hunters* when I was eight," he wrote. (*Microbe Hunters* was written twenty years before the invention of the first influenza vaccine.)

Maher's *Huffington Post* entry also contained several inaccuracies. He argued that polio was on the decline before the vaccine. In fact, in 1943, ten thousand Americans suffered polio; in 1948, twenty-seven thousand; and in 1952, three years before Jonas Salk's polio vaccine, fifty-eight thousand. Maher claimed that the swine flu epidemic was overblown, again unsupported by the facts. Between April 2009, when swine flu entered the United States, and November 2009, when Maher made the claim, forty-seven million Americans had been infected, more than two hundred thousand had been hospitalized, and ten thousand had died, a thousand of whom were children. Finally, Maher wrote that pregnant women didn't need the influenza vaccine—his most dangerous advice. Indeed, pregnant women were seven times more likely to have been hospitalized with swine flu than women of the same age who weren't pregnant.

Maher argued that if most French citizens didn't believe that a swine flu vaccine was necessary, then it must not be necessary. Ironically, in *Religulous*, he didn't extend the same courtesy to those who believe in God. "So, even if a billion people believe something," he said, "it can still be ridiculous."

Finally, when Maher wanted to educate himself about vaccines he called on Barbara Loe Fisher (a media-relations expert), Russell Blaylock (a neurosurgeon), and Jay Gordon (an anti-vaccine pediatrician). Not one of his advisors is an expert in immunology, virology, bacteriology, epidemiology, or toxicology. And not one has ever published a single study on the science of vaccines. Whereas Maher argued that science refuted much of what was stated in biblical teachings, he abandoned science when talking about vaccines.

Jenny McCarthy, Jim Carrey, and Bill Maher have used their celebrity to misinform the public about vaccines, putting children at unnecessary risk. Unfortunately, the phenomenon isn't new. In the 1950s, when epidemiological studies clearly showed that cigarette smoking caused lung cancer, Edward R. Murrow (a broadcast journalist for CBS News) and Arthur Godfrey (a radio and television personality) used their celebrity to argue that the science was contradictory. Both Murrow and Godfrey died of lung cancer.

Although Barbara Loe Fisher doesn't have a medical or scientific background—and has been unable to provide a biological underpinning for her contention that vaccines cause chronic diseases—the media have viewed her as a credible source of information. She has spoken before congressional subcommittees, served on an FDA vaccine advisory panel, and appeared on well-respected news programs such as ABC's *World News Tonight*. Jenny McCarthy, Jim Carrey, J. B. Handley, and Bill Maher, on the other hand, are seen as less reliable, less informed, and less credible by the media. Their voices are heard on anti-vaccine blogs and entertainment television, not at congressional hearings or federal advisory committees. Considered great entertainment, if not somewhat cartoonish, these new anti-vaccine activists have been relatively marginalized. And things would have stayed that way had not two people stepped forward from an unexpected place—two people no one predicted would have ended up on the other side.

Dr. Bernadine Healy was the director of the National Institutes of Health—the single most respected research organization in the United States—during President George H. W. Bush's administration. Although most people have never heard of her, they have certainly heard of NIH. And every time Healy speaks out against vaccines, the appellation "former director of NIH" follows. On May 12, 2008, Sharyl Attkisson interviewed Healy on *CBS Evening News*. Calm, mature, and seemingly well reasoned, Healy did much to discredit those with whom she had previously worked. "This is the time when we do have the opportunity to understand whether or not there are susceptible children [to autism]," she began, "perhaps genetically, perhaps they have a metabolic issue, mitochondrial disorder, immunologic issue, that makes them more susceptible to vaccines. And I think we have the tools today that we didn't have ten years ago, that we didn't have twenty years ago, to try and tease that out." Healy was right that the past decade had witnessed an explosion in techniques likely to reveal the cause or causes of autism. But she was wrong in claiming they hadn't been used. Quite the opposite. During the past decade, several investigators, using the sophisticated techniques mentioned by Healy, have found several genetic defects in children with autism. Others have found structural differences in the brains of autistic children—differences likely to occur in the womb, not following vaccines.

In her interview with Sharyl Attkisson, Healy continued her attack against vaccines, arguing that children might be susceptible "to a component of vaccines, like mercury. I think the government or certain public health officials in the government have been too quick to dismiss the concerns of these families without studying the population that got sick. We should never shy away from science." Healy's rant against public health officials ignored several facts. For one thing, at the time of the *CBS Evening News* interview, the preservative that contained mercury (thimerosal) had been removed from all vaccines given to young infants. For another, far from being unwilling to study whether parents' concerns about mercury were real, public health officials and academic investigators had performed many studies to determine whether mercury in vaccines caused autism or other problems. It didn't. And those studies cost tens of millions of dollars to perform.

Healy concluded her interview by ignoring recent history: "I do not believe that if we identify a particular risk factor that the public would lose faith in vaccines. I think people understand a polio epidemic. I think they understand a measles epidemic. I think they understand congenital rubella. I think they understand diphtheria. Nobody's going to turn their backs on vaccines. I don't believe the truth ever scares people." But some people *have* turned their backs on vaccines. They've turned their backs on MMR vaccine to the point of measles and mumps epidemics. And they've turned their back on Hib vaccine at the cost of their children's lives. The problem isn't that public health officials haven't performed studies or tried to educate the press and public. The problem is that certain people in the media, such as Sharyl Attkisson, Oprah Winfrey, and Larry King, have dismissed these studies, choosing instead to scare the public, presumably to enhance the entertainment value of their shows.

Bernadine Healy's appearances on *CBS Evening News*, her articles in *U.S. News and World Report* (where she is an editor), and her statements in newspapers and magazines have had an effect. But Healy's impact pales in comparison to that of a pediatrician from southern California—a pediatrician who has written a book about vaccines that has influenced a nation.

CHAPTER 10

Dr. Bob

He will look attractive and he will be nice and helpful and he will get a job where he influences a great
God-fearing nation and he will never do an evil thing.
He will just bit-by-little-bit lower standards where they
are important.

—AARON ALTMAN, *BROADCAST NEWS*

R obert Sears is the son of William and Martha Sears. Together,
William, a Harvard-trained pediatrician, and Martha, a regis-
tered nurse and lactation consultant, have authored more than forty
books on pregnancy, birthing, attachment, breastfeeding, nutrition,
sleeping, and discipline—all part of *The Sears Parenting Library*.
Their advice once dominated parenting magazines and the airwaves,
the couple having appeared on *20/20, Donahue, Good Morning
America, Oprah, CBS This Morning*, CNN, *The Today Show*, and
Dateline NBC. Three of their eight children are also doctors, including Jim, who co-hosts the television program *The Doctors*, and
Robert, a pediatrician practicing in southern California.

In October 2007, Robert Sears also published a book. He called
it *The Vaccine Book: Making the Right Decision for Your Child*.
Sears's goal was clear. He wanted to provide what he believed was
a gentler, safer way to vaccinate children—a middle ground for parents who wanted to protect their children but were frightened by so
many shots. Sears has excellent credentials; he received his medical

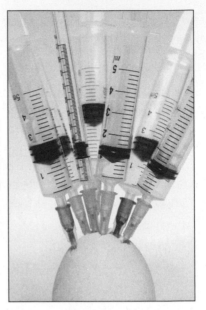

Many parents are concerned
that children are getting too
many vaccines too early.
(Courtesy of David
Gould/Getty Images.)

degree from Georgetown University and his pediatric training from
the Children's Hospital of Los Angeles. Like his father, who prefers
to be called Dr. Bill, Robert Sears prefers Dr. Bob. At the end of his
book, Sears offers a revised schedule he believes is safer than that
recommended by the CDC and AAP. He calls it "Dr. Bob's Alternative Vaccine Schedule." For parents looking for a way to delay,
withhold, separate, or space out vaccines, Dr. Bob's schedule is the
way to go; many parents bring it to their doctor's office and say,
"This is the one I want." Sears's book is so popular, so influential,
and so widely quoted that it deserves a closer look.

"[The alternative schedule] gives live-virus vaccines one at a time so
that a baby's immune system can deal with each disease separately,"
writes Sears. By implying that an infant's immune system is easily
overwhelmed, Sears appeals to a common fear. When Jenny Mc-
Carthy and Jim Carrey led their "Green Our Vaccines" rally in front
of the Capitol, parents marched to the rhythmic chant "Too many
too soon! Too many too soon!" And it's understandable. No rea-

sonable parent can watch a child receive as many as five shots at one time and not worry it's too much. But the fear should be allayed by the science.

Although the number of vaccines given to young children today is more than at any time in history, the immunological challenge from vaccines is lower. A hundred years ago, young children received one vaccine: smallpox. Today, they receive fourteen. But it's not the number of vaccines that counts; it's the number of immunological components contained in vaccines. Smallpox, the largest virus that infects mammals, contains two hundred viral proteins, all of which induce an immune response. Today's fourteen vaccines are made using viral proteins, bacterial proteins, and the complex sugars (polysaccharides) that coat bacteria. Each of these components, like viral proteins in the smallpox vaccine, evokes an immune response. The total number of immunological components in today's fourteen vaccines is about a hundred and sixty, fewer than the two hundred components in the only vaccine given more than a hundred years ago.

Further, Sears fails to consider that vaccines do not significantly increase the immunological challenge that babies encounter and manage every day. In the womb, the unborn child is in a sterile environment. But while passing through the birth canal, the child immediately confronts millions of bacteria. And that's not the end of it; the food that babies eat isn't sterile, nor is the dust they inhale. By the time babies are just a few days old, trillions of bacteria live on the lining of their intestines, nose, throat, and skin. Indeed, people have more bacteria living on the surface of their bodies (a hundred trillion) than they have cells in their bodies (ten trillion). And each bacterium contains between two thousand and six thousand immunological components. Some of these bacteria have the capacity to invade the body and cause harm. To prevent this from happening, every day babies make large quantities of different kinds of antibodies—some of these antibodies pour into the bloodstream (immunoglobulin G), others travel to mucosal surfaces (secretory immunoglobulin A).

Bacteria aren't the only problem. Babies also encounter a variety of viruses that aren't prevented by vaccines—for example, rhinoviruses (which cause the common cold), parainfluenza virus, respiratory syncytial virus, adenovirus, norovirus, astrovirus, echovirus, coxsackie virus, human metapneumovirus, parechovirus, parvovirus, and enterovirus. And, unlike vaccine viruses, which reproduce poorly or not at all, these natural viruses reproduce thousands of times, causing an intense immune response. Arguably, a single infection with a common cold virus poses a much greater immunological challenge than all current vaccines combined. And common viruses occur commonly; healthy children experience as many as six to eight viral infections every year during their first few years of life.

When Sears advised giving live viral vaccines separately, he implied that children have a limited capacity to respond to vaccines. So, how many can they respond to? Do the fourteen vaccines young children receive exceed their immunological capacity? The most thoughtful answer to this question comes from two immunologists at the University of California at San Diego: Mel Cohn and Rod Langman, who study the component of the immune system most capable of protecting against infection: antibodies. Antibodies are made by cells in the body called B cells. Each B cell makes antibodies against only one immunological unit called an epitope. Given the number of B cells in the bloodstream, the average number of epitopes contained in a vaccine, and the rapidity with which a sufficient quantity of antibodies could be made, babies could theoretically respond to about a hundred thousand vaccines at one time.

The model isn't perfect. It assumes that the immune response is static, which it isn't. Every minute new B cells generated in the bone marrow pour into the bloodstream. So, it would be fair to say that at any single point in time a child could theoretically respond to a hundred thousand vaccines. Given that babies are constantly confronted with trillions of bacteria and that each bacterium contains thousands of epitopes, the notion that children could respond to a hundred thousand different vaccines shouldn't be surprising. In a sense, babies are doing that every day. The challenge from vaccines is dwarfed by this natural onslaught.

In 2010, in response to the growing fear of so many vaccines given so early, researchers at the University of Louisville performed a study of more than a thousand children. They found that children who were vaccinated completely and on time were not more likely to suffer neurological problems than children whose parents had chosen to delay vaccination.

Sears advises, "It's probably okay to give the combination MMR booster at age five, when a child's immune system is more mature." Because the MMR vaccine is recommended for children between twelve and fifteen months of age, Sears implies that a baby's immune system isn't mature enough to respond to vaccines. To the contrary, vaccines given in the first year of life induce an excellent immune response. Probably the most dramatic example is the hepatitis B vaccine. Babies born to mothers with hepatitis B virus are not only at high risk of infection, they're also at high risk of chronic liver damage (cirrhosis) and liver cancer. The greatest risk comes at the time of delivery. When passing through the bloody birth canal of an infected mother, babies come in contact with an amazing amount of hepatitis B virus; each milliliter (about one-fifth of a teaspoon) of blood contains about a billion infectious viruses—and birth exposes babies to a lot of blood. So it's no wonder that almost all unimmunized children of infected mothers get infected. But despite the fact that the hepatitis B vaccine is given after exposure, almost all babies are protected. It is rather remarkable that following passage through a birth canal containing literally billions of hepatitis B viruses, a one-day-old baby can mount a protective immune response to a vaccine that contains only twenty micrograms (millionths of a gram) of one highly purified viral protein.

Sears doesn't discourage parents who want to delay vaccines. Under the heading "Delaying Vaccines Until Six Months of Age," he writes, "This choice is one that some parents make, usually for the same reasons as those who wait until two years. They just don't feel as comfortable leaving their child unvaccinated as long. If you've chosen to delay shots, whether it's for six months, one year, or more, you should be aware that your child would not need the entire vaccine series when you do eventually start." Sears implies that a choice to delay vaccines is reasonable. Unfortunately, he fails

to describe the importance of preventing diseases like Hib, pneumococcus, and pertussis, all of which typically appear in the first
year of life and all of which can exact a terrible toll. Most mothers
have antibodies directed against all three of these bacteria and, while
pregnant, pass them to their babies through the placenta. But antibodies from the mother fade, leaving the child vulnerable. Vaccines
against Hib, pneumococcus, and pertussis are given at two, four,
and six months of age so that when the mother's antibodies wear
off, children will have acquired their own protective immunity. Also,
young infants, because they have narrower windpipes, are much
more likely to die from pertussis than older infants. By stating that
a choice to delay vaccines is acceptable, Sears fails to explain why
vaccines are given when they're given.

Sears claims that the most important reason to space out and separate vaccines is to avoid one ingredient: aluminum. "The alternative schedule suggests only one aluminum-containing vaccine at
a time in the infant years," he writes. "By spreading out the shots,
you spread out exposure so infants can process the aluminum
without it reaching toxic levels." Sears explains that "some studies indicate that when too many aluminum-containing vaccines are
given at once, toxic effects occur." In fact, studies show just the
opposite.
      Various preparations of aluminum salts have been used in vac-
cines since the late 1930s. So, the safety of aluminum in vaccines
has been assessed for more than seventy years. Aluminum salts act
as adjuvants, enhancing the immune response. Inclusion of aluminum salts in vaccines that otherwise wouldn't evoke a good immune response makes it possible to reduce the number of doses and
the quantity of immunological components within each dose.
      Although Sears claims that avoiding aluminum-containing vac-
cines is an important way to avoid aluminum, it's not. Aluminum,
the third most abundant element on earth, is everywhere. It's present in the air we breathe, the food we eat, and the water we drink.
The single greatest source of aluminum is food; present naturally in
teas, herbs, and spices, aluminum is also added to leavening agents,

anti-caking agents, emulsifiers, and coloring agents, and is found in pancake mixes, self-rising flours, baking powder, processed cheese, and cornbread. Adults typically ingest 5–10 milligrams (thousandths of a gram) of aluminum every day. Babies are no different; all are exposed to aluminum in breast milk and infant formula. Infants exclusively breast-fed will have ingested ten milligrams of aluminum by six months of age; those fed regular infant formula, thirty milligrams; and those fed soy formula, one hundred and twenty milligrams. All recommended childhood vaccines combined contain four milligrams of aluminum.

Sears is right in stating that aluminum can be toxic, specifically causing brain dysfunction, weakening of the bones, and anemia. But he's wrong in claiming that the small quantities of aluminum in vaccines can be harmful. That's because aluminum has been found to be harmful in only two groups of people: severely premature infants who receive large quantities of aluminum in intravenous fluids, and people on chronic dialysis (for kidney failure) who receive large quantities of aluminum in antacids. In other words, for aluminum to cause harm, a child's kidneys would have to work poorly or not at all *and* the child would have to have received large quantities of aluminum from other sources, such as antacids, which contain more than three hundred milligrams of aluminum per teaspoon.

Other studies are reassuring. Because it's unavoidable, everyone has aluminum circulating in the body, even babies, who have 1–5 nanograms (billionths of a gram) per milliliter of blood. Researchers have studied the quantity of aluminum in blood before and after receipt of aluminum-containing vaccines. No difference. The quantity of aluminum in vaccines is so small and the body eliminates it so quickly (about half of the injected aluminum is completely eliminated in one day) that it is undetectable following vaccination.

To avoid giving more than one aluminum-containing vaccine at a time, Sears advises that children visit their doctors when they are two, three, four, five, six, seven, nine, twelve, fifteen, eighteen, twenty-one, and twenty-four months old (at least twice the number of typical visits). That's a lot of work to avoid a component in vaccines that has never been found to cause harm and is otherwise unavoidable, as-suming that babies ingest breast milk or infant formula.

Like Jenny McCarthy, Sears states that vaccines should be spaced out to avoid a buildup of potentially toxic chemicals. "[The alter-native schedule] gives no more than two vaccines at any one time to limit and spread out exposure to the numerous chemicals so a baby's system can process each more individually," he writes. "Of course, we don't know whether this precaution is necessary, but it's reasonable." Sears describes the chemicals contained in vaccines. In addition to aluminum he lists mercury, formaldehyde, polysorbate 80, monosodium glutamate (MSG), ethylenediaminetetraacetic acid (EDTA), 2-phenoxyethanol, sodium borate, octoxynol, and sodium deoxycholate (all used to promote cell viability, prevent contami-nation, or inactivate bacterial toxins or viruses). He explains that each of these chemicals is potentially harmful: formaldehyde can "cause kidney damage and genetic damage"; monosodium gluta-mate is an "excitotoxin" that "can affect how the brain functions and . . . can damage nerve tissue in a pattern similar to Alzheimer's disease"; 2-phenoxyethanol "may cause reproductive defects and is severely irritating to the eyes and skin"; octoxynol is "used as a spermicide"; and sodium deoxycholate "is harmful if swallowed, in-haled, or absorbed through the skin." For each of these chemicals Sears concludes that the quantity contained in vaccines is "minus-cule," "negligible," or "considered harmless." Then, in an appar-ent contradiction, he advises parents to separate out vaccines to limit exposure and possible harm.

Unfortunately, Sears fails to educate his reader about the impor-tance of quantity—that is, that it's the dose that makes the poison—and that spacing out vaccines to avoid exposure to quantities of chemicals so small that they have no chance of causing harm will accomplish nothing. For example, Sears claims that formaldehyde is a "carcinogen" (cancer-causing agent) but omits the fact that formaldehyde is a natural product: an essential intermediate in the synthesis of amino acids (the building blocks of proteins) and of thymidine and purines (the building blocks of DNA). Everyone has

The "Green Our Vaccines" rally headed by Jenny McCarthy and Jim Carrey expressed the concern that vaccines contained dangerous toxins and chemicals. (Courtesy of Christy Bowe/Corbis.)

about two and one-half micrograms of formaldehyde per milliliter of blood. Therefore, young infants have about ten times more formaldehyde circulating in their bodies than is contained in any vaccine. Further, the quantity of formaldehyde contained in vaccines is at most one six-hundredth of that found to be harmful to animals. It would have been valuable if Sears had informed his readers of these facts rather than scaring them with the notion that formaldehyde in vaccines could cause cancer.

In the preface of his book Sears states, "I want to be clear on something right up front. This is not an *anti-vaccine* book. There are plenty of books out there that overemphasize the potential dangers of vaccines and leave parents even more fearful and confused." But Robert Sears's book isn't what he'd like it to be. Throughout, he implies that vaccines have a high rate of serious side effects, that they aren't adequately tested for safety, that diseases prevented by vaccines aren't that bad, and that pharmaceutical companies misrepresent data. And he makes many claims that are refuted by science.

That's exactly what anti-vaccine books do. Indeed, the themes in Sears's book are the same as those trumpeted in Charles Higgins's *Horrors of Vaccination Exposed and Illustrated*, Lora Little's *Crimes of the Cowpox Ring*, Barbara Loe Fisher's *A Shot in the Dark*, and pamphlets produced by anti-vaccine activists dating back to the mid-1800s.

Sears makes the following arguments:

Vaccines have a high rate of serious side effects. Sears reviews data from the Vaccine Adverse Events Reporting System (VAERS), claiming that between 1991 and 2001 people reported eighteen thousand severe side effects that "resulted in a prolonged hospital stay, a severe life-threatening illness, a permanent disability, or death." Sears concludes that, given the number of doses of vaccines administered during that ten-year period, children had a one in twenty-six hundred chance of suffering serious harm by age twelve. That's a remarkably high rate of serious side effects.

VAERS can, at its best, alert public health officials to the possibility of a serious side effect from a vaccine. VAERS, however, cannot determine whether a vaccine caused a side effect. Only controlled studies can do that. The problem with VAERS is that two groups of people *never* report to it: people who get a vaccine and don't suffer any side effects and people who suffer the same illness as is reported to VAERS but never got the vaccine. This information is critical to determining whether the risk of a possible side effect is greater in the vaccinated group. Sears also fails to address another problem with VAERS: biased reporting. For example, 80 percent of people who reported to VAERS that vaccines caused autism weren't doctors or nurses or nurse practitioners or parents; they were personal-injury lawyers.[10]

The reason that Sears fails to distinguish whether a side effect following a vaccine is actually caused by the vaccine is that, like anti-vaccine activists before him, he simply doesn't believe in coincidence. He writes, "Sometimes infants and children develop medical problems . . . within days or weeks of a vaccination. Although it can be highly suspected that the vaccine was the cause, it can't be proven. I'm sure the truth of the matter is somewhere in between causality and coincidence." Sometimes epidemiological studies find that vaccines cause a problem (e.g., measles-containing vaccine causes a short-lived low platelet count in the blood) and sometimes studies find that they don't (e.g., thimerosal in vaccines doesn't cause autism). In each of these studies a truth has emerged. Sometimes it takes months or years or decades for a truth to emerge. Sometimes it never emerges. But there is one truth: a vaccine either causes a problem or it doesn't. Sears's protests notwithstanding, there's no middle ground between coincidence and causality.

Vaccines aren't adequately tested for safety. Sears writes, "A new medication goes through many years of trials in a select group of people to make sure it is safe. Vaccines, on the other hand, don't receive the same type of in-depth, short-term testing or long-term safety research." In fact, vaccines are tested in larger numbers of people for longer periods of time than any drug. HPV vaccine was tested in thirty thousand women, the conjugate pneumococcal vaccine in forty thousand children, and the current rotavirus vaccines in one hundred and thirty thousand children before licensure; all were tested for more than twenty years. No drug receives this level of scrutiny. And the post-licensure surveillance system for vaccines, specifically the Vaccine Safety DataLink, is a model to detect rare adverse events after a vaccine is licensed. If Vioxx were a vaccine, the fact that it was a rare cause of heart attacks would have been detected far more quickly.

Vaccine-preventable diseases aren't that bad. Sears tells the following story: "A six-month-old unvaccinated infant had a pneumococcal ear infection that spread to the skull bones behind the ear, called mastoiditis. She required surgery and IV [intravenous] antibiotics. Afterward, I asked the parents if they regretted their decision not to vaccinate. They said no. They were both well-educated professionals, had done a lot of reading on this issue, and still felt comfortable with their decision." Sears implies that, because the child survived, pneumococcal infections aren't really that bad (or that surgery isn't really that bad). It doesn't always work out that way. Every year many children suffer pneumococcal pneumonia, bloodstream infections, and meningitis. And those who don't die from meningitis are often left blind, deaf, or mentally disabled. For example, in 2001 Shannon Peterson of Minnesota decided not to give her two children the pneumococcal vaccine. Both suffered severe pneumococcal infections. Her five-year-old son survived; her six-year-old daughter didn't. "I can't tell parents enough the importance of vaccines," said Peterson. "I hope that no one else has to hold their child when they die." Sears could have told a story like this one. It certainly happens often enough. But he didn't. Instead he referred in glowing terms to the parents of a child who needlessly suffered mastoiditis. The truth is these parents had made a terrible decision for their child—one that could have killed her.

Vaccines contain dangerous ingredients. In the mid-1800s, anti-vaccine activists claimed that vaccines contained the "poison of adders, the blood, entrails, and excretions of bats, toads and suckling whelps." When, a hundred and fifty years later, Jenny McCarthy said that she wanted the ether and anti-freeze removed from vaccines, she had carried forward the centuries-old tradition of claiming that vaccines contain ingredients that aren't there. Vaccines of old didn't contain products derived from adders, bats, toads, or whelps; today's don't contain ether or anti-freeze.

Sears, like McCarthy, claims that vaccines contain phantom ingredients. He writes that some vaccines are made using serum obtained from calves before they're born. That's true. Then he takes an illogical step, raising the specter of mad-cow disease. "All animal and human tissues are carefully screened for all known infectious diseases," he writes. "Some vaccine critics are still worried, however, that there may be other viruses or infectious agents called 'prions' . . . that are much smaller than viruses and that we don't yet know how to screen for." Proteinaceous infectious particles (prions) cause mad-cow disease, a progressive dementia that often results in death. Mad-cow disease swept through the British beef industry in the 1980s, killing one hundred and sixty British citizens; with stricter regulations, the disease has been eliminated. It would have been helpful if Sears had mentioned several reassuring facts: prions grow in the nervous system, not in cells used to make viral vaccines; prions have never been found to contaminate serum obtained from calves before they're born; mad-cow disease isn't a problem in New Zealand (where calf serum is obtained); and children receiving vaccines during the mad-cow epidemic weren't at increased risk of mad-cow disease. Although most parents probably never considered mad-cow disease before they read his book, Sears concludes, "If exposure to animal tissues worries you, you may want to choose the brand that doesn't use cow extract."

Sears's fear of phantom vaccine ingredients didn't end with prions. Regarding the MMR vaccine, he wrote, "The measles and mumps vaccines are nourished for years in a culture of chicken embryo cells [that contain] human albumin, a protein filtered out of human blood units." Sears is correct in stating that MMR is stabilized using human serum albumin. And he's right in stating that it's a blood protein. But the human albumin in MMR isn't made from human blood; it's made using recombinant DNA technology. Human blood is never part of the process. Sears's misstatements are a far cry from claims that vaccines contain the blood of bats and toads—just not far enough.

Pharmaceutical companies misrepresent data. Sears writes, "Twenty years ago a group of doctors from the CDC, several U.S. medical centers, and two pharmaceutical companies—GlaxoSmithKline and Merck—undertook the task of determining just how common the hep[atitis] B infection was in infants and children. If they found that hep[atitis] B was very common in kids, it would make sense to begin vaccination of all newborns. The consensus of the researchers was that [thousands of] infants and children were being infected with this virus each year." Sears didn't believe it. Taking a closer look, he found only "about 360 cases reported in kids from birth through age nine each year." Sears implied that the CDC, GlaxoSmithKline, and Merck had misled the public.

It's not hard to appeal to the public's distrust of government and pharmaceutical companies. Lora Little did it in *Crimes of the Cowpox Ring* and Barbara Loe Fisher in *A Shot in the Dark*. But like Fisher's and Little's claims, Sears's aren't supported by the facts. Before the hepatitis B vaccine was recommended for babies in 1991, every year about sixteen thousand children less than ten years old were infected with the virus. Given that many hepatitis B virus infections occur without symptoms—and are not reported to the CDC—this estimate is probably low.

On January 20, 1961, during his inaugural address, President John F. Kennedy said, "Ask not what your country can do for you. Ask what you can do for your country." Twenty years later, Ronald Reagan, during a debate with President Jimmy Carter, asked, "Are you better off now than you were four years ago?" Both men understood the prevailing mood. Kennedy had appealed to a sense of community, sending thousands of young people into programs like the Peace Corps and Volunteers in Service to America (VISTA); he asked Americans to see themselves as part of something greater, to take responsibility for something greater. Reagan appealed to the "Me Generation"; now it was time for me to get mine.

A parallel can be drawn with vaccines. On February 2, 2009, a show titled "The Polio Crusade" aired on public television's *American Experience*. The program described a polio outbreak in the summer of 1950 that devastated the town of Wytheville, Virginia. And it told the story of America's efforts to make the first polio vaccine. It's a remarkable program. Throughout the documentary are heard the voices of Americans sixty years ago, and they reveal a heart-warming sense of community. People saw polio as a shared tragedy, giving millions of dollars to the March of Dimes to make a vaccine. And they gave more than their money; thousands of community organizers volunteered to conduct the largest field trial of a vaccine ever performed—one that included about two million children. When it was over—when a polio vaccine emerged that eliminated the disease from the Western Hemisphere—Americans were proud. They felt that they, more than anyone else, had developed the vaccine. Individuals saw themselves as part of a group—a public that cared about public health. It was this sentiment that John F. Kennedy so deftly touched during his inaugural address.

Sears, like Reagan before him, is appealing to a generation that doesn't consider a larger cooperative—an immunological commons. Toward the end of his book, under the heading "Is It Your Social

Responsibility to Vaccinate Your Kids?" he writes, "This is one of the most controversial aspects of the vaccine debate. Obviously, the more kids who are vaccinated, the better our country is protected and the less likely it is that any child will die from a disease. Some parents, however, aren't willing to risk the very rare side effects of vaccines, so they choose to skip the shots. Their children benefit from herd immunity—the protection of all the vaccinated kids around them—without risking the vaccines themselves." Sears then asks the critical question. "Is this selfish? Perhaps. But as parents you have to decide. Are you supposed to make decisions that are good for the country as a whole? Or do you base your decisions on what's best for your own child as an individual? Can we fault parents for putting their own child's health ahead of other kids around him?" In another section of the book, Sears doesn't hide the deceit. Regarding parents who are afraid of the MMR vaccine, he writes, "I also warn them not to share their fears with their neighbors, because if too many people avoid the MMR, we'll likely see the diseases increase significantly." In other words, hide in the herd, but don't tell the herd you're hiding. Otherwise, outbreaks will ensue. Sears's advice was prescient. Within a year of the publication of his book, the United States suffered a measles epidemic that was larger than anything experienced in more than a decade. (It was an outbreak fueled by the unfounded fear that MMR vaccine caused autism—a fear that Sears fails to allay in his book.)

Now that herd immunity has broken down, Sears's position that one should think only of oneself no longer works. Unfortunately, his book contains many examples of this philosophy:

- "In truth, tetanus is not an infant disease," he writes. "Also, diphtheria is virtually non-existent in the United States. So you could create a logical argument that a baby could skip the tetanus and diphtheria shots for a few years and be just fine." These statements are inaccurate. First: tetanus *is* a disease of infants. A cursory look at any textbook of infectious diseases provides grim pictures of newborns suffering severe muscle spasms and breathing difficulties from tetanus; that's why it's called the "disease of the seventh day." Second: the casual advice that one can simply wait to get a diphtheria vaccine ignores history. Between 1990 and 1993, when public health programs were disrupted in the Russian Federation (states newly independent from the Soviet Union), a hundred and fifty thousand people suffered diphtheria and five thousand died, mostly children. In the absence of vaccination, such an outbreak could happen in the United States just as easily.

- "[Polio] doesn't occur in our country," writes Sears, "so the risk is zero for all age groups." Although polio has been eliminated from the United States, it hasn't been eliminated from the world. Four countries—India, Nigeria, Pakistan, and Afghanistan— have never interrupted polio transmission; and children in twenty-three other countries still suffer the disease. Because international travel is common, and because most people who are contagious aren't sick, it is likely that poliovirus walks into the United States every year. Children whose parents follow Sears's advice will be particularly vulnerable when an outbreak occurs or when they travel overseas.

- "Hib is a bad bug," writes Sears. "Fortunately, it's also a rare bug, so rare that I haven't seen a single case in ten years. Since the disease is so rare, Hib isn't the most critical vaccine." As Sears knows, Hib is rare because of the Hib vaccine. And if we stop using the vaccine, Hib will be back. Which is exactly what has happened. Sears's book was published in October 2007. The following year, outbreaks of Hib meningitis occurred in Minnesota and Pennsylvania. All these outbreaks centered on children whose parents had chosen not to vaccinate them; four died from their infections.

Robert Sears peers out from the back cover of his book with an open, caring expression, exuding a kind of California calm. No doubt he wants to do the right thing; no doubt he is trying to find some middle ground between parental anxiety about getting vaccines and physician anxiety about not giving them; no doubt he believes he is on the side of his fellow physicians. Describing his "alternative schedule," Sears writes, "I have put together a vaccine schedule that gets children fully vaccinated, but does so in a way that minimizes the theoretical risks of vaccines. It's the best of both worlds of disease prevention and safe vaccination." But Sears's schedule is ill-founded. And rather than calming parents with science that exonerates vaccines, he caters to their fears by offering a schedule that has no chance of making vaccines safer and will only increase the time during which children are susceptible to infections that can kill them. It's the worst of both worlds.

Although Sears is probably well meaning, one has to question the hubris of a man who decides to create his own vaccine schedule— someone who claims his schedule is better and safer than that recommended by the CDC and AAP. It's all the more amazing when one considers that Robert Sears has never published a paper on vaccine science; never reviewed a vaccine license application; never participated in the creation, testing, or monitoring of a vaccine; and never developed an expertise in any field that intersects with vaccines— specifically, virology, immunology, epidemiology, toxicology, microbiology, molecular biology, or statistics. Yet he believes he can sit down at his desk and come up with a better schedule. And parents trust him. Oddly, they trust him *because* he doesn't have an expertise in vaccine science—an expertise that would likely have inspired the CDC, AAP, FDA, professional medical organizations, or vaccine makers to seek his advice.

One final irony. For a new vaccine to be added to the schedule, the FDA requires concomitant-use studies. Pharmaceutical companies must show that a new vaccine doesn't interfere with the immunity or safety of existing vaccines and that existing vaccines don't interfere with the new vaccine. Only then can a vaccine become part of the schedule. Dr. Bob's schedule, on the other hand, is completely untested—never reviewed by the FDA, CDC, or AAP to make sure it's as safe and effective as the existing schedule. It is remarkable how little Sears thinks of the enormous amount of testing that goes into creating the current schedule.

Sears isn't alone.

On January 12, 2010, Dr. Mehmet Oz, host of the popular *The Dr. Oz Show*, told interviewer Joy Behar what he thought about the influenza vaccine.

> BEHAR: There's a rumor that your kids did not get flu shots or swine flu shots. Is that right?
> OZ: That's true. They did not.
> BEHAR: Do you believe in them for the kids or what?
> OZ: No. I would have vaccinated my kids but you know I—I'm in a happy marriage and my wife makes most of the important decisions as most couples have in their lives.

Given their relative training, one would have imagined that Oz, not his wife, would have made the decision. Mehmet Oz graduated from Harvard University in 1982 and obtained a joint MD and MBA degree from the University of Pennsylvania School of Medicine and the Wharton School in 1986. Since then, he's climbed the ranks to become a professor of cardiac surgery at Columbia University. His wife, Lisa, has no background in science or medicine. Rather, Lisa Oz is guided by the beliefs of Mikao Usui, who, after three weeks of fasting and meditation on Mount Kurama in Japan, claimed he had been given the power to heal through his palms—called reiki. Lisa Oz isn't just a follower of Usui, she's a reiki master.

The four Oz children weren't among the hundreds of thousands hospitalized or the hundreds killed by swine flu in 2009. But they could have been. And the influenza vaccine would have prevented it. No scientific evidence supports palm healing as a method to treat or prevent influenza.

Oz's disdain for vaccines didn't end on *The Joy Behar Show*. In December 2009, Oz and co-author Michael Roizen published *YOU: Having a Baby*, a book that promotes Dr. Bob's Alternative Vaccine Schedule. Oz and Roizen wrote, "One of the most highly charged conflicts revolves around an issue that comes up just moments after your baby is born: to vaccinate or not to vaccinate? That, indeed, is one heck of a question." Like Sears, Oz and Roizen misinformed their readers on several counts:

Mehmet Oz, host of *The Dr. Oz Show*, often dispenses anti-vaccine advice. Shown here with wife Lisa at *Time* magazine's 100 Most Influential People Gala, May 8, 2008. (Courtesy of Scott McDermott/Corbis.)

- Regarding the polio vaccine, they wrote, "There's no doubt that polio vaccine . . . causes polio in 1 in 1 million to 2 million," failing to mention that the only polio vaccine available today in the United States is inactivated and, therefore, incapable of causing polio.
- Regarding the influenza vaccine, they wrote, "Pregnant women should avoid getting the influenza vaccine in their first trimester." Instead of the vaccine, they suggest that "you can boost your immune system during the winter by taking 2,000 IU [International Units] of vitamin D daily." Pregnant women are much more likely to be hospitalized and killed by influenza than nonpregnant women of the same age. That's why they're asked to receive the influenza vaccine if they're pregnant during influenza season. The vaccine, not vitamin D, induces specific immunity to the virus.
- Regarding the rotavirus vaccine, they wrote, "A prior version of this vaccine was withdrawn from the market in 1999 because it was linked to a severe condition known as intussusception, a

blockage or twisting of the intestine. A new vaccine, released in 2006, has been associated with even more cases of intussusception . . . than the first version, prompting an FDA notification in 2007. We recommend that you opt out of this one until more data are available." Oz and Roizen should have read the FDA notification a little more carefully. If they had, they would have seen that the FDA stated that all cases of intussusception following rotavirus vaccine may have occurred by chance alone. Further, one year before *YOU: Having a Baby* was published, the CDC found the risk of intussusception was the same in children who did or didn't receive the rotavirus vaccine; parents no longer have to wait for data.

Robert Sears and Mehmet Oz have followed in the footsteps of anti-vaccine activists before them, claiming to inform parents about vaccines while in fact misinforming them. Their popularity has only widened the gap between some parents and their pediatricians.

So how does one solve the problem of the growing rift between parents who are concerned about the safety of vaccines and doctors who are worried about the reemergence of infectious diseases? The solution may not be easy; but it's there.

CHAPTER 11

Trust

Leave the gun. Take the cannolies.
—PETER CLEMENZA, *THE GODFATHER*

W e've reached a tipping point. Children are suffering and
dying because some parents are more frightened by vaccines
than by the diseases they prevent. It's time to put an end to this. Several solutions have been proposed. The first would be effective but
is too awful to imagine; the second, given the history of the American legal system, will never happen; the third, while possible, would
require a sea change in our culture.

In 1994, Robert Chen, then head of immunization safety at the Centers for Disease Control and Prevention, created a graph titled "The
Natural History of an Immunization Program." Chen described
what happens when vaccines are used for a long time, partitioning
the public's reaction into distinct phases.

In the first phase, people are afraid of infections. In the 1940s,
parents readily accepted the diphtheria and pertussis vaccines because diphtheria and pertussis commonly killed young children; and
the tetanus vaccine because many people died of tetanus, especially
during World Wars I and II. In the 1950s, parents rushed to get the
polio vaccine because they saw what polio could do; everyone knew
someone who had been paralyzed or killed by the virus. In the
1960s, parents gave their children measles, mumps, and rubella vaccines because they had witnessed firsthand the devastation wrought

by those diseases: pneumonia and encephalitis from measles, deafness from mumps, and severe birth defects from rubella. During this phase of Chen's graph, immunization rates rise.

In the next phase, as vaccines cause a dramatic reduction in disease, vaccines become a victim of their own success. Now the focus is on vaccine side effects, real or imagined. Immunization rates plateau.

In the next phase, as fear of vaccines continues to rise, immunization rates fall. And preventable diseases increase. It's in this phase that America now finds itself. When Chen showed this graph to colleagues at the CDC, he used statistics to support his argument. And, like most thoughtful scientists, he remained dispassionate, referring to children as numbers on a graph. But there was emotion in those numbers. ("Statistics are people with their eyes wiped dry," said former Surgeon General Julius Richmond.)

The last and most disturbing phase of Chen's graph offers a solution to the problem posed by unvaccinated children. In this phase, the incidence of preventable deaths becomes so high that parents again seek solace in vaccines. Immunization rates rise. In a more perfect world, we would never get to this part of Robert Chen's graph. We would learn from history—learn from the smallpox deaths in England in the late 1800s following widespread anti-vaccine activity, learn from pertussis deaths in England and Japan in the mid-1970s following unfounded fears that the vaccine caused brain damage, learn from measles deaths in England and Ireland in the late 1990s caused by the false notion that MMR caused autism, and learn from bacterial meningitis deaths in Minnesota and Pennsylvania in 2009 caused by the fear that children were getting too many vaccines.

Although renewed fears of fatal infections caused by the reemergence of these diseases would undoubtedly increase vaccination rates, the price is far too great.

Another solution to the problem of unvaccinated children would be to eliminate religious and philosophical exemptions.

Religious exemptions would be impossible to eliminate. That's because parents have been letting their children die in the name of religion for decades, without consequence. And these children have been denied treatments that *would* have saved their lives, not vaccines that *might* have saved their lives. For example:

- In the late 1890s and early 1900s, many children of Christian Scientists died from diphtheria, even though diphtheria antitoxin was widely available. Christian Science healers were unrepentant, one stating, "We do not feel bound to the laws of hygiene, but to the laws of God." Several parents were charged with manslaughter, none successfully.
- In 1937, Edward Whitney, a widowed insurance salesman, left his ten-year-old daughter, Aubrey, in the care of her aunt in Chicago. Aubrey was a diabetic. The aunt, a Christian Scientist, took Aubrey to her practitioner, William Rubert, who immediately took her off insulin. On December 10, 1937, Aubrey Whitney died in a diabetic coma; Rubert wasn't held accountable for her death. Twenty-two years later, Edward Whitney walked into Rubert's office, pulled out a 32-caliber handgun, and shot him at point-blank range.
- In 1951, Cora Sutherland, a fifty-year-old Christian Scientist who taught shorthand at Van Nuys High School in Los Angeles, argued successfully that she should be exempt from the periodic X-rays required by her school to detect tuberculosis. Three years later, in March 1954, she died of tuberculosis, but not before exposing thousands of students. The health department petitioned the board of education to eliminate religious exemptions, without success.
- In 1955, seven-year-old David Cornelius became ill; his parents, Edward and Anne Cornelius, took him to a doctor who diagnosed diabetes and started insulin. Later, a Christian Science clinician stopped the insulin, causing David to die in a diabetic coma. The district attorney indicted Edward and Anne Cornelius for involuntary manslaughter, but dropped the charges when a senior church official persuaded him that the Corneliuses "had sincerely believed that they could save their son through prayer."

- In 1967, Lisa Sheridan, the five-year-old daughter of Dorothy Sheridan, contracted strep throat. For the next three weeks Lisa found it harder and harder to breathe. Dorothy, a Christian Scientist, prayed but to no avail; Lisa Sheridan died of pneumonia on March 18, 1967. At autopsy, the pathologist found a quart of pus in Lisa's chest that had collapsed her lung—pus that could have been removed easily had Dorothy sought medical attention. Sheridan was convicted of manslaughter and sentenced to five years in jail. The church, frightened by the verdict, issued a scathing rebuke: "We must not yield to the mesmeric claims of medicine by calling a doctor and being forced to worship a false God." Christian Science officials successfully lobbied the Department of Health and Human Services to exempt faith healers from prosecution. In 1974, when the federal exemption was made, eleven states already had a religious exemption statute; ten years later, all fifty states and the District of Columbia had it.

- In 1977, Matthew Swan, the second child of Rita and Douglas Swan, had a high fever. The Swans asked their Christian Science practitioner, Jeanne Laitner, to treat him. Laitner complied; sitting next to Matthew's crib she said, "Matthew, God is your life. God didn't make disease, and disease is unreal." Matthew continued to scream in pain. On July 7, Matthew Swan was pronounced dead from bacterial meningitis. Unlike other Christian Science parents, Rita Swan saw the death of her son as a wake-up call. She founded Children's Health Care Is a Legal Duty (CHILD), an organization devoted to changing religious exemption laws.

But the neglect continued:

- On March 9, 1984, Shauntay Walker died of bacterial meningitis. Her mother, Laurie, a Christian Scientist, had kept her home for seventeen days. At the time of death, Shauntay, who was five years old, weighed only twenty-nine pounds. In 1990, Walker was convicted of manslaughter, but the conviction was overturned with the help of her lawyer, Warren Christopher, who would later become Bill Clinton's secretary of state.

- On April 8, 1986, Robyn Twitchell—the two-year-old son of David and Ginger Twitchell—died of a bowel obstruction. David and Ginger had graduated from a Christian Science college in Missouri. After the bowel obstruction ruptured, Robyn vomited stool and portions of his bowel. He died in his father's arms. At trial, Dr. Burton Harris, chief of surgery at Boston's Floating Hospital for Infants and Children, testified, "It's beyond comprehension that the parents of a child who's vomiting stool wouldn't seek medical help." The Twitchells were found guilty; the verdict was overturned on appeal.

- On June 5, 1988, twelve-year-old Ashley King died of bone cancer. The only child of John and Catherine King, Ashley lay at home for months without medical care. At the time of her death, the tumor was the size of a watermelon; her hemoglobin level was incompatible with life, and she was covered with bedsores. John and Catherine King each pleaded no contest to one charge of reckless endangerment: a misdemeanor. They were sentenced to three years' probation.

- On May 9, 1989, eleven-year-old Ian McKown—the son of Kathleen McKown—died in a diabetic coma. Doctors testified that insulin given even two hours before his death could have saved his life. The police officer called to the house said that the child was so emaciated that "he didn't even look human." Kathleen McKown was protected from prosecution by Minnesota's religious exemption law.

Despite deaths at the hands of faith healers, religious exemptions have remained intact, causing prosecutors either to decline to file criminal charges or to lose in court. Only three states—Massachusetts, Hawaii, and Maryland—have repealed their religious exemption health laws; the rest continue to offer protection to parents who medically neglect their children in the name of God.

The notion that U.S. courts would eliminate religious exemptions to vaccination, when they haven't eliminated religious exemptions to lifesaving medicines, is fanciful.

Philosophical exemptions, which have become increasingly more popular, would also be difficult to eliminate.

In the 1990s, philosophical exemptions were available in only a handful of states; now, they're available in twenty-one. Alan Hinman, the CDC official interviewed for *Vaccine Roulette* who actively promoted state mandates in the 1970s, doesn't see any hope of eliminating philosophical exemptions. "I don't think that one would win the battle in the legislature on getting rid of philosophical or personal belief exemptions," he said. "Looking at the trajectory of our society over the last several years, I find it hard to imagine. If anything, we're going the other way." Walter Orenstein thinks philosophical exemptions at the very least should be much more difficult to obtain. "I believe that a decision *not* to vaccinate is of equal gravitas to [the decision] *to* vaccinate," he said. "And there should be a procedure whereby people have to read information, understand information, and sign that they understand the risks they are putting that child through. Right now, in some places, it's a hell of a lot easier to get an exemption than to get your child vaccinated." Orenstein sees his proposal as of value only for those who choose what he calls "exemptions of convenience." "For people who are adamantly opposed to vaccines," he says, "I don't think this will make that much difference."

Another solution would be for the medical community to respond more directly to the threat of decreasing immunization rates.

Recently, hospital administrators have been mandating influenza vaccines given yearly for healthcare providers. Regarding influenza, certain facts are unassailable: people sickened by influenza come into the hospital, healthcare providers can spread influenza virus from one patient to another, patients who catch influenza in the hospital can suffer severe and fatal illness, and hospitals with higher rates of immunization among healthcare providers have lower rates of influenza. Despite these facts, influenza vaccination rates among healthcare providers have been woeful—hovering around 40 percent. So, in the name of patient safety, hospital administrators are doing something about it.

In 2009, eight hospitals in the United States mandated influenza vaccine for their employees. Some took a softer approach: if a healthcare provider refused vaccination, administrators required a surgical mask to be worn throughout the day. Others took a harder line. At the Children's Hospital of Philadelphia healthcare providers who refused influenza vaccination were given two weeks of unpaid leave to think about it. If they still refused, they were fired. As a consequence, immunization rates among healthcare providers at the hospital rose from 35 percent in 2000 to 99.9 percent in 2010. Administrators at Children's Hospital knew they were responsible for a vulnerable population; so they stood up for them.

Doctors are also doing something that decades ago would have been unthinkable: they're refusing to see parents who won't vaccinate their children.

For doctors, it's a lose-lose situation. Doctors who refuse to care for unimmunized children are sending a strong message. They're saying vaccines are so important that they cannot be asked to withhold them. The problem with this approach is that by refusing to see unimmunized children, doctors lose any chance of convincing parents of the value of vaccines; worse still, these children will likely remain unimmunized and vulnerable. On the other hand, if doctors continue to see unimmunized children, they're sending a tacit message that it's an acceptable choice. And it's not a choice that parents are making for their child only; it's a choice they're making for other children, including those in the doctor's waiting room. The measles outbreaks in 2008 are a perfect example of how parents' choices for their children affected others. When unimmunized children developed fever and a rash, parents brought them to their pediatricians' offices, where other children, some too young

to be immunized, were exposed. Doctors' offices became epicenters
of measles transmission. Now, doctors are asking: who will stand
up for children in our waiting rooms if not us?

Brad Dyer, a pediatrician in Lionville, Pennsylvania, has written
a vaccine policy that is posted throughout his office. "We call it the
manifesto," he says. The document, titled "The Importance of Immunizing Children," reads, in part, as follows:

*We firmly believe in the effectiveness of vaccines to prevent seri-
ous illnesses and save lives.*

We firmly believe in the safety of vaccines.

*We firmly believe that all children and young adults should re-
ceive all of the recommended vaccines according to the schedule
published by the Centers for Disease Control and Prevention and
the American Academy of Pediatrics.*

*We firmly believe based on all available literature, evidence,
and current studies that vaccines do not cause autism or other
developmental disabilities.*

*Furthermore, by not vaccinating your child you are taking self-
ish advantage of thousands of others who do vaccinate their children, which decreases the likelihood that your child will contract
one of these diseases. We feel such an attitude to be self-centered
and unacceptable.*

*We are making you aware of these facts not to scare you or co-
erce you, but to emphasize the importance of vaccinating your
child. We recognize that the choice may be an emotional one for
some parents. We will do everything we can to convince you that
vaccinating according to the schedule is the right thing to do.
Please be advised, however, that delaying or "breaking up the vaccine" to give one or two at a time over two or more visits goes
against expert recommendations and can put your child at risk for
serious illness (or even death) and goes against our medical advice.*

*Finally, if you should absolutely refuse to vaccinate your child
despite all our efforts, we will ask you to find another healthcare
provider who shares your views. We do not keep a list of such
providers nor would we recommend any such physician.*

Following the posting of the policy, only a handful of parents left Dyer's practice. "Parents have said, 'Thank you for saying that. We feel much better about it,'" he says.

Unfortunately, nothing will change if the push to vaccinate comes only from doctors, vaccine advocates, public health officials, and hospital administrators. Some parents will always view these groups as biased; and it hasn't been hard for anti-vaccine groups to appeal to the sentiment that they can't be trusted.

When parents choose to vaccinate their children, one element is critical to the decision: trust. A choice not to vaccinate is a choice not to trust those who research, manufacture, license, recommend, promote, and administer vaccines—specifically, the government, pharmaceutical companies, and doctors. If we are to again believe that vaccines are safer than the diseases they prevent, we're going to have to trust those responsible for them. This isn't going to be easy.

The CDC has been a particularly useful target. The mere mention of the term *the government* conjures up images of labyrinthine red tape and career bureaucrats who don't care about anything except their pensions. But anyone who spends time with the people at the CDC responsible for vaccines will come away with a far different impression. People like Walter Orenstein, a pediatrician and former director of the National Immunization Program, who early in his career worked to eradicate smallpox, caring for one of the last cases in India. Or Anne Schuchat, an internist and director of the National Center for Immunization and Respiratory Diseases, whose compassion toward those sickened and, in some cases, killed by the novel H1N1 virus was apparent at every CDC press briefing. Or Larry Pickering, a pediatrician, infectious diseases specialist, and executive secretary of the Advisory Committee on Immunization Practices, who dedicated his early career to understanding how infections in child-care centers could best be eliminated. Or Nancy Messonier, a tireless advocate for children who suffer meningococcal infections. John Salamone, a parent advocate who worked closely with CDC officials during his fight to change polio-vaccine policy, was impressed by what he saw: "They were all incredibly professional, all very caring, all wanting to do the right thing."

Eventually, we are going to have to appreciate that CDC officials aren't against us—they are us. Those involved with vaccines include doctors and scientists who are also parents, aunts, uncles, and grandparents. "We're human," says Walter Orenstein. "We have children. And we use the same vaccines in our own children as we recommend for anybody else."

Another group targeted by anti-vaccine groups is vaccine advocates. People like Deborah Wexler, who visits housing projects to talk about the importance of vaccines, and provides free vaccines to Southeast Asian refugees, all through an organization she founded in St. Paul, Minnesota. Or David Tayloe, former president of the American Academy of Pediatrics, who has served his North Carolina community as a practicing pediatrician for more than four decades. Or Amy Pisani, who studied at Gallaudet University for the deaf before becoming executive director of Every Child by Two, an organization that provides vaccine services to low-income families.

Despite the efforts of these advocates for children, journalists with a direct link to the anti-vaccine movement have tried to discredit them. For example, on July 25, 2008, Sharyl Attkisson aired a segment on the *CBS Evening News with Katie Couric*. (Attkisson also maintains an anti-vaccine blog site.) Attkisson had discovered what she believed really motivated people like Amy Pisani: money. "Every Child by Two," said Attkisson, "a group that promotes early immunization for all children, admits the group takes money from the vaccine industry, too, but wouldn't tell us how much. A spokesman [Amy Pisani] told us, 'There are simply no conflicts to be unearthed.' But guess who has been listed as the group's treasurer?—an official from Wyeth and a paid advisor to big pharmaceutical clients." Attkisson's implication wasn't subtle. Every Child by Two was a sham, merely fronting for the interests of Big Pharma. But Attkisson's guilt-by-association reporting lacks one critical element: quid pro quo. Where is the evidence that Every Child by Two, an organization founded by the wives of a former U.S. president and a former senator (Rosalynn Carter and Betty Bumpers)—with minuscule salaries, a shoestring budget, and a tiny office in Washington, D.C.—is promoting vaccines for any reason other than its belief that they are lifesaving? And where is the evidence that the money that it receives from pharmaceutical companies is used for anything other than outreach to underprivileged children? Attkisson's report was the worst kind of journalism: damning by association. Pharmaceutical companies provide unrestricted educational grants to groups like Every Child by Two all the time. The key part of those grants is the word *unrestricted*. Once given, the company has no say in how that money is spent.

Attkisson didn't stop with Every Child by Two. She also attacked the American Academy of Pediatrics. "The vaccine industry gives millions to the American Academy of Pediatrics," she said, "for conferences, grants, medical education classes, even to help pay to build their headquarters. The totals are kept secret, but public documents reveal bits and pieces; $342,000 was given to the academy by Wyeth, maker of the pneumococcal vaccine for a community grant program; $433,000 was contributed to the academy by Merck, the same year the academy endorsed Merck's HPV vaccine. Another top donor, Sanofi-Aventis, maker of seventeen vaccines and a new five-in-one-combo shot just added to the vaccine schedule last month." Attkisson implied that pharmaceutical companies bought the AAP's support of vaccines. But the AAP's recommendations are based on a careful review of safety and efficacy data. Is Attkisson saying that vaccines wouldn't be recommended by the AAP unless pharmaceutical companies greased its palms? It's a fantastic accusation. Isn't it possible that the AAP promotes vaccines for the same reason that the FDA licenses them and the CDC recommends them—that vaccines save lives? Is there any evidence that members of the AAP or Every Child by Two, after looking at data on vaccine safety or effectiveness, have realized that the vaccines fell short of what is necessary to help children? And, if so, then turned a blind eye? Have representatives from any organization designed to further the health and welfare of children—goals that are directly in line with the promotion of vaccines—ever said to themselves, "Sure, the data on vaccine safety don't look very good here, but we'll ignore those studies because we receive unrestricted educational grants?" Because that's what Attkisson was implying—that the relationship is bad for our nation's children. But if she is going to make that accusation, she will need more evidence than the fact that an association exists.

Attkisson's attacks weren't unique. Pharmaceutical companies are often a target for cynicism and distrust. Indeed, few industries are more reviled. And, to some extent, it's understandable. In order to sell their products, pharmaceutical companies have occasionally acted aggressively, unethically, and even illegally. Pfizer's $2.3 billion settlement with the government for providing false information about Bextra, a pain medication, and Viagra, a well-known potency product, are probably the most extreme examples of how far companies will go to sell their products. So, by extension, it hasn't been hard to tarnish the reputations of people who speak on behalf of vaccines. Lora Little, Barbara Loe Fisher, Jenny McCarthy, J. B. Handley, Jim Carrey, Bill Maher, and other anti-vaccine activists have consistently railed against the unholy alliance between vaccine makers and those who promote vaccines. The implication is clear— any association with vaccine makers is unacceptable. The media buy it; congressional committees buy it; parents buy it. And, once bought, the implication effectively eliminates a lot of expertise from the debate—leaving people like Robert Sears, a man with no published experience on vaccine science or vaccine safety, to do a lot of the talking.

There is, however, one problem with the image of the evil vaccine maker. At no time in history has a pharmaceutical company ever engaged in illicit marketing practices for vaccines. Not once. And it's not because government regulators aren't watching. They're watching. Indeed, companies that make vaccines are often taken to task for improper marketing of drugs. It's unclear why this is true. Maybe it's because drugs are much more lucrative than vaccines. Or maybe it's because company employees who make and promote vaccines are more likely than those who promote drugs to have a public health background, so they see vaccines as a public service. Whatever the explanation, it seems reasonable to suggest that if we are going to eliminate from the debate anyone who has come in contact with vaccine makers, we should at least get our money's worth. It would be compelling if at some point in the two-hundred-year history of vaccine manufacture one shred of evidence existed that contact with a vaccine maker led to information that was misleading or incorrect. Sharyl Attkisson, in her hit piece against Every Child by Two and the AAP, never once showed how her unsubstantiated allegation of undue influence by vaccine makers led to anything other than better-educated physicians and healthier children.

At the heart of Attkisson's attacks is the notion that the government, pharmaceutical companies, and doctors are part of a conspiracy. During his tenure at the CDC, Walter Orenstein was a target. "One of the most frustrating things about the anti-vaccine activity is the conspiracy theories," recalled Orenstein. "Scientists often argue with each other. We say, 'Your studies are poor; you've misinterpreted your data; you didn't consider this study.' But no one ever thinks in a scientific debate that if you say something with which someone else disagrees that you're lying. I think one of the frustrations I find here is that if we say anything that doesn't confirm [the anti-vaccine groups'] preconceived notions, we're automatically labeled as liars."

Conspiracy theories lie at the heart of the anti-vaccine movement, claiming that the pharmaceutical industry, using undue influence, causes eighty thousand practicing pediatricians and family physicians to lie about vaccine safety. "This season's fashion in conspiracy theories—for those out of the loop of enlightenment—concerns health," wrote David Aaronovitch, author of *Voodoo Histories: The Role of Conspiracy Theory in Modern History*, in the *Wall Street Journal*. "The Web sites, marginal cable shows, and radio phone-ins are full of tales about how Big Pharma and Bad Government are deliberately spreading diseases or manufacturing scares in order to sell us expensive drugs, gull us into dangerous vaccinations or just simply to create an atmosphere of panic which will allow 'them' to take over. We live in an age of conspiracies, or rather, we are more aware of conspiracies than we used to be. It is better to think that someone is in charge of everything than that the world is more often prey to accidents, madness, and coincidence. That's why movies are full of dastardly but brilliant plotters, and hardly anything happens by chance."

Probably the best example of the casual conspiracy theorist can be found after the decision by the VICP special masters that thimerosal in vaccines didn't cause autism. Rebecca Estepp, a mother of a child with autism, said, "I was disappointed but not surprised. Vaccine court is a system where government attorneys defend a government program using government-funded science decided by government judges. I don't think these children had much of a chance." Estepp's notion of a government conspiring to deny her child compensation through the VICP wasn't supported by the facts. First: most of the research that failed to show that thimerosal caused autism was funded by academia, not the U.S. government. Second: the "government judges" who had denied her claim that vaccines caused autism had awarded almost $2 billion to petitioners since 1989 for other claims, primarily in cases where the evidence wasn't clear. For thimerosal, the evidence was abundantly clear. But what was most striking about Estepp's comments was that virtually every major news organization, including the *New York Times*, carried them unchallenged.

Conspiracy theories proliferate because of one shared characteristic between vaccine makers and the government: lack of a definable face. Companies are portrayed as caring only about money to be made, not lives to be saved. The media never tell, and the public never hears, stories of people at pharmaceutical companies who are directly involved in vaccine research and development.

Penny Heaton is a pediatrician who trained in infectious diseases at the University of Louisville. Then she joined the Epidemiology Intelligence Service at the CDC investigating diarrheal diseases in Africa. Between 1997 and 1999, Heaton, working in Africa, saw children die from overwhelming dehydration, mostly caused by ro-

Drs. Penny Heaton (center of photo), Kathy Neuzil, and Abraham Victor Obeng
Hodgson in Northern Ghana during rotavirus vaccine field trials. (Courtesy of
Penny Heaton.)

tavirus. In 1999, she decided to do something about it, joining
Merck to lead its rotavirus vaccine program. It was a daunting task.
At the end of a pre-licensure study that took four years to complete,
involved eleven countries, included seventy thousand children, and
cost $350 million, Heaton called together two hundred people at
her company. She began by showing a map of the world. "This is
what the world looks like now," she said, pointing to hundreds of
small black dots concentrated in Asia, Africa, and Latin America.
"Each of these dots represents a thousand deaths a year from rotavirus." Then Heaton showed a map without any black dots.
"Now," she said, pointing to the clean map, "we have the technology in hand to eliminate deaths from this disease." Then she wept.
She stood in front of two hundred people with her head down and
her shoulders shaking, thinking of those children in Africa. This
isn't an image that anyone has of a pharmaceutical company. And
it's unlikely that anyone ever will. Because no matter how much

companies make people like Penny Heaton available to the press and the public, no matter how much they try to provide a compassionate face to their industry, one belief will remain: vaccine manufacture is an enterprise that makes money and makes money only. Heaton later helped to provide Merck's vaccine to several African countries.

If we are to get past the constant barrage of misinformation based on mistrust, we have to set aside our cynicism about those who test, license, recommend, produce, and promote vaccines. Only then will we survive this detour—a detour that has caused far too many children to suffer needlessly.

EPILOGUE

We must not be enemies. Though passion may have
strained it must not break our bonds of affection . . .
when again touched, as surely they will be, by the
better angels of our nature.

—ABRAHAM LINCOLN

Maybe the tide will turn only when parents start to speak up.
Remarkably, that's happening.

On December 19, 2008, a program titled "Ruining It for the
Rest of Us" aired on National Public Radio. It centered on the
measles epidemic in San Diego started by the seven-year-old unvaccinated boy who caught the disease in Switzerland. When he
got home, after walking among a group of football fans at the air-
port on their way to the Pro Bowl in Hawaii, he went to a Whole
Foods Market, Trader Joe's, Chuck E. Cheese, a swim school for
babies, his elementary school, and his pediatrician's office. During
his travels as many as 980 people were exposed to measles. Once
alerted to the situation, the health department resorted to quarantine. Anyone judged to have had direct contact with the boy—
or to have shared a space that he had occupied in the previous two
hours—was asked to stay home for twenty-one days. The boy was
placed in protective isolation on a military base. When the outbreak was over, twelve children had contracted measles and sixty
had been quarantined.

"Ruining It for the Rest of Us" contained voices typically found
in an anti-vaccine story:

There was an anti-vaccine mother, Sybil Carlson, who, like the mother of the seven-year-old boy who had started the outbreak, had chosen not to vaccinate her children. She said she no longer trusted doctors and the heavy-handed manner in which they tried to force her to vaccinate.

There was the moderator, Susan Burton, who, in reference to vaccine ingredients, said, "I'm pretty sympathetic to this stuff. When Sybil says that aluminum, a vaccine additive, is a known neurotoxin, I'm right there with her." (The measles vaccine doesn't contain aluminum.)

And there was Robert Sears: "There's a great quote in *Star Trek* where Spock says, 'The needs of the many outweigh the needs of the few or the one.' There's no better way that you can look at that than with vaccines. Spock was saying this in a movie while he was dying to save the Starship *Enterprise*. And you know, would Spock's mom have said he made a good decision because she had to lose her son over it? It's a big decision and I think parents are ultimately going to do what they think is best for the one instead of what they view as best for the many."

But unlike most programs on vaccines, "Ruining It for the Rest of Us" featured parents who were angry that they had to suffer the choices of others. For example, there was Hillary Chambers, whose daughter, Finley, had been quarantined for three weeks. Burton asked her how she had dealt with the quarantine. "It was so sudden I was scared," she said. "What was I going to do for the next three weeks? My husband and I both work." (Before vaccines, mothers typically stayed home with their children during quarantines. Vaccines made it easier for women to enter the workplace.) Then Burton asked Hillary about her reaction to the outbreak. "I was really angry," she said. "It impacted families financially, emotionally, on so many different levels. So I was mad. And wanted to know how this happened." Finally, Burton asked whether Hillary sympathized with parents' decisions not to vaccinate. "I understand that it's scary and that getting vaccinated is a leap of faith," she said. "The battle starts when one person's choices affect other people. And that's not being a responsible member of this community."

Another parent, Meagan Campbell, was also interviewed. Unlike Hillary, Meagan hadn't been inconvenienced by quarantine; she'd been forced to watch her son suffer and almost die from the disease. While waiting in his pediatrician's office, Meagan's son was exposed to the boy who had contracted measles in Switzerland. Because he was only ten months old, he was too young to have received the MMR vaccine. Meagan described her ordeal: "The rash started getting worse and spreading down over his body. On Saturday my parents came down from Los Angeles and took one look at him and said your kid has the measles—because they were from a generation that knows what it looks like. At that point my son had the full rash all the way to his toes and we would never put him down, even for a nap. I thought as long as someone was holding him then we knew his heart was beating. There were moments when I didn't think he was going to make it because the fever just wasn't letting up—106 degree fever and this rash that made my son look like an alien almost. And I wondered when it was over whether he was going to be the same boy that he was before. I was a mom that has probably taken fifty pictures of him every day since the day he was born. But there are not pictures of this until he was finally better because it wasn't my son and I never want to remember him looking that way."

Burton asked Meagan about her reaction to the outbreak. "I just wondered [about] this family that had brought this outbreak into San Diego," she said. "What were they thinking? Did they feel for us at all? Did they feel bad about it? I have very close friends who don't vaccinate their children. And it's just something that we can't talk about. We get too angry and we can barely speak. I feel like if I were to engage in the conversation we might not be friends anymore."

Burton asked, "Do you think it should be a choice—that people should be able to opt out of the measles vaccine?" Meagan thought before answering. "Yes," she replied, "but they have to live on an island: their own little infectious diseases island. Don't go to the same doctors as the rest of us. Don't go to the same schools. Don't go to the same stores. Live on an island somewhere if that's the choice you want to make."

Celina Yarkin and
family on their
Vashon Island farm,
2010. (Courtesy of
Celina Yarkin.)

Hillary Chambers and Meagan Campbell aren't the only parents speaking out.

Celina Yarkin lives on Vashon Island, Washington, a community with dreadfully low immunization rates. She's the mother of three children: Adrianna, age nine; Eleanor, six; and Madeline, four.

Yarkin's story is similar to most on the island. Born and raised in Seattle, she attended Antioch and then Evergreen College, where she studied visual arts. In 1996 she joined the Peace Corps, teaching English in Guinea-Bissau, Africa, before returning to live on a collective in Albuquerque, New Mexico. In 2001, when her eldest daughter was nine months old, she moved to Vashon Island, where she farms. "We're a market farm," she says, "so we grow 20–30 different kinds of vegetables [and then] bring them to the farmers' market. We do two markets a week."

Soon after arriving, Yarkin knew that something was different about Vashon Island. "When we moved here the BBC and the Boston public broadcasting channel were doing a production on vaccines," she recalled. "They came to Vashon Island to show a community that had really high unvaccinated rates. I ended up being in that film because we had goats and a goat trailer. So they used me as a shot of . . . a typical parent. I said to them that I'd like to talk about my views on vaccines because I think they're really important. But they didn't want to talk to me. They wanted to talk to people who were really resistant."

Yarkin was concerned about the large number of unvaccinated children in her community. "Most of my concern is that we're vulnerable to outbreaks. It's like when you go to other countries and you can see things that are broken and you say 'Oh gosh, if we could just do it differently.' This is something in my community that I think is dysfunctional. It concerns the safety of our kids. I thought this is something I'm going to take a stand on; I want to talk about it; I want to work through it with people—to try and punch back at all of that information on the Internet that everyone's getting about autism and vaccines and the dangers."

To start, Yarkin wanted to know exactly how many children in her community were unimmunized. So she made a few phone calls. In February 2009, Yarkin headed a meeting on the island that included Walter Orenstein; Zachary Miller, an infectious diseases expert from Seattle; public health officials from the state and county; and the principals from both the local high school and elementary school: "The result of the meeting was that we didn't know what the vaccination rates on the island were; but we needed to figure that out. It's just amazing that people were so ready for someone to step forward from the community." Yarkin also traded e-mails with William Foege, who designed the strategy that eventually eliminated smallpox from the world. Foege is a resident of Vashon Island, too.

After months of reviewing charts from physicians' offices and public health clinics, Yarkin had an answer. Pertussis and chickenpox immunization rates on the island were well below those necessary to prevent outbreaks. Worse: the estimates were probably optimistic. "The numbers we have are skewed in favor of vaccines because those who go to alternative practitioners or are homeschoolers aren't counted," she said.

To increase immunization rates, Yarkin is constructing displays to educate parents at elementary schools and health clinics. "And when the displays go up," she says, "I will get the local newspaper to draw people's attention to it." She knows it isn't going to be easy. "There's like a parenting code. Do you tell other parents that you disagree with what they're doing? No. You don't interfere with other families' decisions about how they're raising their kids." But Yarkin sees herself in those who are choosing not to vaccinate. "I was one of those moms who questioned vaccines based on the autism scare, [which] showed me my vulnerabilities to unsound scientific information. Going forward, I hope to be less vulnerable." And to make others less vulnerable. "I took this on because I believe people want good information. And they want to do the right thing for their families and their community. They'll change their behavior if they believe the information is sound."

Yarkin knows that her efforts could alienate friends. "I'm okay with being in a controversy. It's okay with me if people disagree and back away from me. It's something that I've had to really work through. But I don't think I'll lose any real friends over it. And I won't be insulted or hurt if people don't do what I want them to do." In the end, she remains optimistic: "I feel I have a chance to really change the course of events."

Yarkin's outspokenness is fueled by the fear of what might happen if children in her community remain unvaccinated. More typically, parent activism is fueled by events that have already affected their children. Such is the case with the mother of Julieanna Flint of Waconia, Minnesota.

In January 2008, when Julieanna was fifteen months old, she had an episode of vomiting and slight fever. Her mother, Brendalee, thought little of it. The next morning Julieanna's temperature rose. So Brendalee took her daughter to Heidi Wuerger, her family doctor. "It sounds like flu," she was told. "Give her Tylenol." Brendalee took her daughter home, gave her plenty of liquids, and washed her with a cool, damp washcloth. But the fever didn't go away. As the day wore on, Julieanna refused to eat or drink.

The next day, she woke up screaming with a temperature of 104 degrees. "She couldn't say 'Help me,' but her eyes were begging me to do something," recalled Brendalee, who rushed her to the emergency room at Ridgeview Medical Center, where doctors performed a spinal tap. But instead of the fluid being clear (as it should be), it was cloudy; and instead of containing no white blood cells (cells the body uses to fight infection), it contained 145,000. "Your daughter is seriously ill," Dr. Wuerger told Brendalee. "We need to transport her to Children's [Hospital in Minneapolis], now!"

At Children's Hospital, doctors gave Julieanna antibiotics to treat her bacterial meningitis. Her fever decreased. But the next day Julieanna had a seizure, the first of many. That same day, the cause of her meningitis was revealed: Hib. At that point, a Hib vaccine had been available for twenty years, and Julieanna had received every dose of it. Yet she still suffered the disease. Later, doctors figured out why: Julieanna had an inability to make antibodies. So, even though she'd gotten the vaccine, she'd never developed an immune response to it.

Julieanna worsened. During an MRI scan, doctors found a massive collection of pus surrounding her brain. "We're doing everything we can to save your daughter's life," a doctor told Brendalee. "I still remember walking her to the surgery room and giving her to the doctor," recalled Brendalee. "I didn't know if I would see her again." Brendalee called a priest to administer last rites. "Through this holy anointing may the Lord in his love and mercy help you with the grace of the Holy Spirit," said the priest, marking Julieanna's forehead with oil in the shape of a cross. "May the Lord who frees you from sin save you and raise you up."

Julieanna slowly began to recover. As a result of her meningitis, she had to relearn how to swallow, crawl, walk, and talk. "It was like having a newborn again," recalled Flint. "I would rub her throat for swallowing and rub her cheeks for chewing. She couldn't crawl. She could scream and that was about it." But she was alive. Although Brendalee didn't know it, the percentage of children in Minnesota whose parents had refused to give them Hib vaccine had increased sixfold during the previous few years. Because Julieanna couldn't make antibodies, she was particularly vulnerable.

Brendalee and Julieanna
Flint, December 16, 2009.
(Courtesy of Andy King.)

In April 2009, Brendalee and Julieanna Flint traveled to Washington, D.C., to speak to congressional staffers about the importance of vaccines. "Parents need to understand that when they choose not to vaccinate, they are making a decision for other people's children as well," said Brendalee. "Someone else chose Julieanna's path. It doesn't seem fair that someone like Jenny McCarthy can reach so many people while my little girl has no voice."

Following the tragedy of September 11, 2001, there was a moment when we all stood still and looked at each other. No longer individuals, we were part of a whole. Personal interests were irrelevant. We were united in our grief. One.

Then the moment was gone, dissolved in a cloud of lawsuits, finger-pointing, partisanship, and blame. But, although fleeting, it had been there. And if we can recapture it—recapture the feeling that we are all in this together, all part of a large immunological cooperative—the growing tragedy of children dying from preventable infections can be avoided. We can do this. It's in us: the better angels of our nature.

NOTES

Prologue

ix **Parents choose not to vaccinate their children:** G. L. Freed, S. J. Clark, A. T. Butchart, et al. "Parental Vaccine Safety Concerns in 2009," *Pediatrics* 125 (2010): 1–6; P. J. Smith, S. G. Humiston, Z. Zhao, et al., "Association Between Delayed or Refused Vaccination Doses and Timely Vaccination Coverage," abstract presented at the Pediatric Academic Society's annual meeting, Vancouver, British Columbia, May 4, 2010.

ix **Percentage of unvaccinated children has more than doubled:** S. B. Omer, W.K.Y. Pan, N. A. Halsey, et al., "Nonmedical Exemptions to School Immunization Requirements: Secular Trends and Association of State Policies with Pertussis Incidence," *Journal of the American Medical Association* 296 (2006): 1757–1763.

ix **Doctors refuse to see unvaccinated children:** E. A. Flanagan-Klygis, L. Sharp, and J. E. Frader, "Dismissing the Family Who Refuses Vaccines: A Study of Pediatrician Attitudes," *Archives of Pediatric and Adolescent Medicine* 159 (2005): 929–934.

Introduction

xi **Robert Bazell:** *NBC Nightly News*, February 17, 2009.

xi **Minnesota outbreak:** Centers for Disease Control and Prevention, "Invasive *Haemophilus Influenzae* Type B Disease in Five Young Children—Minnesota, 2008," *Morbidity and Mortality Weekly Report* 58 (2008): 1–3.

xi **Parent of child with Hib:** *NBC Nightly News*, February 17, 2009.

xii **Hib outbreaks:** D. Sapatkin, "A Fatal Link in Vaccine Shortage," *Philadelphia Inquirer*, April 1, 2009.

xii **Hib disease:** Centers for Disease Control and Prevention, "Invasive *Haemophilus Influenzae* Type B Disease in Five Young Children—Minnesota, 2008," *Morbidity and Mortality Weekly Report* 58 (2008): 1–3.

xii **Whooping cough before vaccine:** J. Cherry, P. Brunnel, and G. Golden, "Report of the Task Force on Pertussis and Pertussis Immunization," *Pediatrics* 81 (suppl.) (1988): 933–984.

xii **Whooping cough today:** Centers for Disease Control and Prevention, "Preventing Tetanus, Diphtheria, and Pertussis Among Adolescents: Use of Tetanus Toxoid, Reduced Diphtheria Toxoid and Acellular Pertussis Vaccines," *Morbidity and Mortality Weekly Report* 55 (2006): 1–42.

xii **Vashon Island:** K. Rietberg, C. DeBolt, N. Heimann, et al., "School Exemption Data and Geocoding as Tools for Assessing Relationships of Pertussis Clusters to High Immunization Exemption Rates," http://cdc .confex.com/cdc/nic2002/recordingredirect.cgi/id/1446; "Public Health Skepticism," November 30, 2002, http://medpundit.blogspot.com/2002 _11_24_medpundit_archive.html.

xiii **Whooping cough symptoms:** K. M. Edwards and M. D. Decker, "Pertussis Vaccines," in *Vaccines*, 5th ed., eds. S. A. Plotkin, W. A. Orenstein, and P. A. Offit (London: Elsevier/Saunders, 2008).

xiii **El Sobrante:** H. K. Lee, "Whooping Cough Outbreak Closes Private School in El Sobrante," http://sfgate.com/cgibin/article.cgi?f=/c/a/2008/ 05/10/BAN310JQVL. DTL; "East Bay School Opens After Whooping Cough Outbreak," http://cbs5.com/local/whooping.cough.pertussis.2 .722073.html.

xiii **Rudolf Steiner:** R. Steiner, *Fundamentals of Anthroposophical Medicine* (Toronto: Mercury Press, 1986).

xiii **"Karmic development":** OpenWaldorf.com, http://www.openwaldorf .com/health.html.

xiii **Nationwide whooping cough outbreaks:** Centers for Disease Control and Prevention, "Pertussis Outbreak in an Amish Community—Kent County, Delaware, September 2004–February 2005," *Morbidity and Mortality Weekly Report* 55 (2006): 817–821; Illinois Department of Public Health, "Pertussis Outbreak Prompts Public Health Warning: At Least 69 Sickened in Four Chicago-Area Counties," http://www.idph.state.il.us/ public.press04/6.18.04.htm, June 18, 2004; Centers for Disease Control and Prevention, "Use of Mass Tdap Vaccination to Control an Outbreak of Pertussis in a High School—Cook County, Illinois, September 2006– January 2007," *Morbidity and Mortality Weekly Report* 57 (2008): 796– 799; Mississippi State Department of Health, "State Health Officials, CDC Studying Pertussis Outbreak in Mississippi," http://msdh.ms.gov/ msdhsite/index.cfm/23,5373,279.html, August 30, 2007; "Pertussis Outbreak Reaches 500 Cases Statewide," Arizona Department of Health Services, http://www.asdhs.gov/news/2005-all/pout.htm; Centers for Disease Control and Prevention, "School-Associated Pertussis Outbreak— Yavapai County, Arizona, September 2002–February 2003," *Journal of the American Medical Association* 291 (2004): 1952–1954; "Pertussis Outbreak Points to the Importance of Vaccination," Oregon Department of Human Services News Release, http://www.oregon.gov/DHS/news/ 2008news/2008-0117.shtml; Centers for Disease Control and Prevention, "Pertussis Outbreak—Vermont, 1996," *Morbidity and Mortality Weekly Report* 46 (1997): 822–826; "CDC Reported Highest Number of Whoop-

ing Cough Cases in Nearly Forty Years," http://www.medicalnewstoday
.com/articles/15220.php; Centers for Disease Control and Prevention,
"Pertussis—United States, 1997–2000," *Morbidity and Mortality Weekly
Report* 51 (2002): 73–76.

xiv **Delaware pertussis outbreak:** Centers for Disease Control and Preven-
tion, "Pertussis Outbreak in an Amish Community—Kent County,
Delaware, September 2004–February 2005," *Morbidity and Mortality
Weekly Report* 55 (2006): 817–821.

xiv **Ashland, Oregon:** S. Robinson, A. Timmons, and L. Duncan, "School Exemptions and Disease Risk in Ashland, Oregon," http://www.co.jackson
.or.us/files/school%20exemptions%20and%20disease%20risk%20-%
20final.pdf.

xiv **Indiana measles outbreak:** A. A. Parker, W. Staggs, G. H. Dayan, et al.,
"Implications of a 2005 Measles Outbreak in Indiana for Sustained Elim-
ination of Measles in the United States," *New England Journal of Med-
icine* 355 (2006): 447–455; Centers for Disease Control and Prevention,
"Import-Associated Measles Outbreak—Indiana, May–June 2005,"
Morbidity and Mortality Weekly Report 54 (2005): 1073–1075.

xv **Measles complications:** P. M. Strebel, M. J. Papania, G. H. Dayan, and
N. A. Halsey, "Measles Vaccine," in *Vaccines*, 5th ed., eds. S. A. Plotkin,
W. A. Orenstein, and P. A. Offit (London: Elsevier/Saunders, 2008).

xv **CDC warnings:** Centers for Disease Control and Prevention, "Update:
Measles—United States, January–July 2008," *Morbidity and Mortality
Weekly Report* 57 (2008): 893–896.

xv **San Diego outbreak:** Centers for Disease Control and Prevention, "Out-
break of Measles—San Diego, California, January–February 2008,"
Morbidity and Mortality Weekly Report 57 (2008): 203–206; R. G. Lin
II, "Rise in Measles Cases Worries Health Officials," *Los Angeles Times*,
May 2, 2008.

xv **California vaccination rates:** R. G. Lin II, "California Schools' Risks Rise
as Vaccinations Drop," *Los Angeles Times*, March 29, 2009.

xvi **Nationwide measles outbreak, 2008:** Centers for Disease Control and
Prevention, "Update: Measles—United States, January–July 2008," *Morbidity and Mortality Weekly Report* 57 (2008): 893–896.

xvi **International measles epidemics:** Centers for Disease Control and Prevention, "Measles Outbreaks in the United States: Public Health Pre-
paredness, Control and Response in Healthcare Settings and the
Community," http://www2a.cdc.gov/HAN/ArchiveSys/ViewMstV.asp
?AlertNum=00273; R. G. Lin II, "Rise in Measles Cases Worries Health
Officials," *Los Angeles Times*, May 2, 2008; S. van den Hof, M.A.E.
Conyn-van Spaendonck, and J. E. van Steenbergen, "Measles Epidemic
in the Netherlands, 1999–2000," *Journal of Infectious Diseases* 186
(2002): 1483–1486.

xvi **Decline in measles immunization rates in certain communities:** D. A.
Salmon, M. Haber, E. J. Gangarosa, et al., "Health Consequences of

Religious and Philosophical Exemptions from Immunization Laws: Individual and Societal Risks of Measles," *Journal of the American Medical Association* 281 (1999): 47–53; D. R. Feikin, D. C. Lezotte, R. F. Hamman, et al., "Individual and Community Risks of Measles and Pertussis Associated with Personal Exemptions to Immunization," *Journal of the American Medical Association* 284 (2000): 3145–3150.

xvi **Mumps outbreak:** Centers for Disease Control and Prevention, "Update: Mumps Outbreak—New York and New Jersey, June 2009–January 2010," *Morbidity and Mortality Weekly Report* 59 (2010): 125–129.

xvii **Herd immunity:** P.E.M. Fine and K. Mulholland, "Community Immunity," in *Vaccines*, 5th ed., eds. S. A. Plotkin, W. A. Orenstein, and P. A. Offit (London: Elsevier/Saunders, 2008).

xviii **Tammam Aloudat and Walter Orenstein regarding polio in Africa and the United States:** Quoted in "Heightened Awareness, Renewed Commitment Needed to Eradicate Polio," *Infectious Diseases in Children*, June 2009.

xviii **Vaccine-preventable diseases in 1900s:** S. A. Plotkin, W. A. Orenstein, and P. A. Offit, eds., *Vaccines*, 5th ed. (London: Elsevier/Saunders, 2008).

1. The Birth of Fear

1 **Wiseman's life:** Frederick Wiseman, U.S. documentary filmmaker, http://museum.tv/archives/etv/W.htmlW/wisemanfred/wisemanfred.htm.

1 *Titicut Follies:* Zipporah Films, Bridgewater Film Company, 1967.

2 **Reviews of *Titicut Follies*:** Benson and Anderson, *Reality Fictions*, 55; V. Canby, "*Titicut Follies* Observes Life in Modern Bedlam," *New York Times*, October 4, 1967; "Trials: Banned in Massachusetts," *Time*, January 19, 1968.

2 **Theatrical poster for *Titicut Follies*:** http://www.zipporah.com.

2 **Judge's ruling on *Titicut Follies*:** Benson and Anderson, *Reality Fictions*, xxiii.

2 *Titicut Follies* **ban:** Ibid., xxiv.

2 *DPT: Vaccine Roulette*: WRC-TV, Washington, D.C., April 19, 1982. All documentary-related quotes and graphics subsequently cited in this chapter were obtained from this source.

6 **Doctors' response to *DPT: Vaccine Roulette*:** E. R. González, "TV Report on DTP Galvanizes US Pediatricians," *Journal of the American Medical Association* 248 (1982): 12–22.

6 **Parents' response to *DPT: Vaccine Roulette*:** Ibid; Hearings before the Subcommittee on Investigations and General Oversight of the Committee on Labor and Human Resources, U.S. Senate, Ninety-Seventh Congress, Second Session to Examine Adverse Drug Reactions from Immunization, Federal Efforts in Preventive Medicine, and Characteristics of Certain Diseases, May 7, 1982, Written Testimony of Vincent A. Fulginiti, M.D., Chairman, Committee on Infectious Diseases, American Academy of Pediatrics.

6 **Switchboard melting:** Author interview with Alan Hinman, December 7, 2009.

7 **Kathi Williams regarding son's reaction to DTP:** First International Conference on Vaccination, National Vaccine Information Center, 1997.

7 **Jeff Schwartz regarding daughter's reaction to DTP:** Quoted in Allen, *Vaccine*, 252–253.

8 **Barbara Loe Fisher regarding son's reaction to DTP:** B. L. Fisher, "In the Wake of Vaccines," *Mothering*, September 1, 2004.

9 **Paula Hawkins, Lea Thompson, and Dan Mica:** Allen, *Vaccine*, 254–255.

10 **Hawkins's testimony:** Hearings before the Subcommittee on Investigations and General Oversight of the Committee on Labor and Human Resources, U.S. Senate, Ninety-Seventh Congress, Second Session to Examine Adverse Drug Reactions from Immunization, Federal Efforts in Preventive Medicine, and Characteristics of Certain Diseases, May 7, 1982.

10 **Williams's testimony:** Ibid.

10 **AAP at Hawkins hearing:** Ibid.

11 **Edward Mortimer and vaccine harm:** Ibid.

11 **Thompson's career:** Lea Thompson, "Dateline NBC Correspondent," http://www.msnbc.msn.com/id/3949442.

12 **Thompson recalls *DPT: Vaccine Roulette*:** First International Public Conference on Vaccination, National Vaccine Information Center, 1997.

2. This England

14 **John Wilson at Royal Society:** B. Deer, "The Vanishing Victims," *Sunday Times* (London), November 1, 1998.

14 **Wilson describes cases of vaccine harm:** M. Kulenkampff, J. S. Schwartzman, and J. Wilson, "Neurological Complications of Pertussis Inoculation," *Archives of Disease in Childhood* 49 (1974): 46–49.

14 **Madsen paper:** T. Madsen, "Vaccination Against Whooping Cough," *Journal of the American Medical Association* 101 (1933): 187–188.

14 **Werne and Garrow paper:** J. Werne and I. Garrow, "Fatal Anaphylactic Shock: Occurrence in Identical Twins Following Second Injection of Diphtheria Toxoid and Pertussis Antigen," *Journal of the American Medical Association* 131 (1946): 730–735.

14 **Byers and Moll paper:** R. K. Byers and F. C. Moll, "Encephalopathies Following Prophylactic Pertussis Vaccine," *Pediatrics* 1 (1948): 437–457.

14 **Ström paper, 1960:** J. Ström, "Is Universal Vaccination Against Pertussis Always Justified?" *British Medical Journal* 2 (1960): 1184–1186.

14 **Ström paper, 1967:** J. Ström, "Further Experience of Reactions, Especially of a Cerebral Nature, in Conjunction with Triple Vaccination: A Study Based on Vaccinations in Sweden 1959–65," *British Medical Journal* 4 (1967): 320–323.

14 Wilson's background: B. Deer, "The Vanishing Victims," *Sunday Times* (London), November 1, 1998.

15 British television program: Ibid.

15 "Risks Concealed": "Whooping Cough Vaccine Risks Concealed, Say Victims' Parents," *Sunday Times* (London), September 22, 1974.

15 "Vaccine Call": "Vaccine Call Is Attacked," *Sunday Times* (London), June 26, 1977.

15 "Should Be Abandoned": "Whooping Cough Vaccine 'Should Be Abandoned,'" *The Times* (London), February 13, 1978.

15 "New Campaign": P. Healy, "New Campaign to Win State Help for the Vaccine-Damaged," *The Times* (London), January 15, 1980.

15 "Shot in the Dark": G. Stewart, "Dangers of a Shot in the Dark," *The Times* (London), September 8, 1982.

15 "Boy's Brain Damaged": "Boy's Brain Damaged in Vaccine Experiment," *Sunday Times* (London), May 30, 1976.

15 Dick paper: G. Dick, "Letter: Whooping-Cough Vaccine," *British Medical Journal* 4 (1975): 161.

15 Stewart editorial: G. Stewart, "Dangers of a Shot in the Dark," *The Times* (London), September 8, 1982.

15 David Kerridge: Quoted in "Whooping Cough Vaccine 'Should Be Abandoned,'" *The Times* (London), February 13, 1978.

15 Health officials concerned: "Family Doctors Split over Whooping Cough Vaccine Dangers," *Sunday Times* (London), March 13, 1977.

16 Editorial describes outbreak of fear: "Whooping-Cough Vaccination," *British Medical Journal* 4 (1975): 186–187.

16 Pertussis in England in late 1970s: J. P. Baker, "The Pertussis Vaccine Controversy in Great Britain, 1974–1986," *Vaccine* 21 (2003): 4003–4010; E. R. González, "TV Report on DTP Galvanizes US Pediatricians," *Journal of the American Medical Association* 248 (1982): 12–22.

16 Salisbury on pertussis epidemic: Author interview with David Salisbury, November 30, 2009.

16 Pertussis in Japan: E. J. Gangarosa, A. M. Galazka, C. R. Wolfe, et al., "Impact of Anti-Vaccine Movements on Pertussis Control: The Untold Story," *The Lancet* 351 (1998): 356–361.

16 Rosemary Fox: B. Deer, "The Vanishing Victims," *Sunday Times* (London), November 1, 1998.

17 Cherry regarding pertussis outbreak in England: Author interview with James Cherry, November 27, 2009.

17 Survey of general practitioners: O. Gillie, "Family Doctors Split over Whooping Cough Vaccine Dangers," *Sunday Times* (London), March 13, 1977.

17 Pertussis deaths in the United Kingdom: J. D. Cherry, "The Epidemiology of Pertussis and Pertussis Immunization in the United Kingdom and the United States: A Comparative Study," *Current Problems in Pediatrics* 14 (1984): 7–77.

17　Pertussis deaths classified as SIDS: Author interview with James Cherry, November 27, 2009; A. Nicoll and A. Gardner, "Whooping Cough and Unrecognized Postperinatal Mortality," *Archives of Disease in Childhood* 63 (1988): 41–47.

18　Miller study: D. L. Miller, E. M. Ross, R. Alderslade, et al., "Pertussis Immunization and Serious Acute Neurological Illness in Children," *British Medical Journal* 282 (1981): 1595–1599; D. Miller, N. Madge, J. Diamond, et al., "Pertussis Immunization and Serious Acute Neurological Illness in Children," *British Medical Journal* 307 (1993): 1171–1176.

18　Lea Thompson and lawyers: *DPT: Vaccine Roulette*, WRC-TV, Washington, D.C., April 19, 1982.

18　Increased number of DTP lawsuits: A. R. Hinman, "DTP Vaccine Litigation," *American Journal of Diseases in Children* 140 (1986): 528–530.

19　Tyler White: *White v. Wyeth Laboratories*, 40 Ohio St. 3d 390.

19　Michelle Graham: *Graham v. Wyeth Laboratories*, 666 F. Supp. 1483

19　Michelle Graham award: Allen, *Vaccine*, 272.

19　Melanie Tom award: Ibid.

19　Other DTP awards: W. A. Check, "How to Remedy Possible Harm to a Few Persons from Vaccines That Could Benefit Entire Society?" *Journal of the American Medical Association* 252 (1984): 2942–2946.

19　Increase in monetary value of lawsuits: A. R. Hinman, "DTP Vaccine Litigation," *American Journal of Diseases in Children* 140 (1986): 528–530.

19　Increased cost of lawsuits: Ibid.

19　Decreased number of companies making DTP vaccine: Letter from H. M. Meyer, director of the Bureau of Biologics to Dr. Bruce Dull of the Centers for Disease Control, December 1, 1975. This letter was generously provided by Dr. Bruce Weninger, CDC.

19　Three companies making DTP in 1982: A. R. Hinman, "DTP Vaccine Litigation," *American Journal of Diseases in Children* 140 (1986): 528–530.

19　Wyeth drops out: Allen, *Vaccine*, 283.

19　Connaught drops out: M. H. Smith, "National Childhood Vaccine Injury Compensation Act," *Pediatrics* 82 (1988): 264–269.

20　James Mason testifies: Testimony before the Subcommittee on Health and the Environment, Committee on Energy and Commerce, U.S. House of Representatives, December 19, 1984.

20　AAP emergency meeting: "DTP Vaccine Shortage," Summary of February 12, 1985, meeting hosted by the AAP; summary generously provided by Dr. Bruce Weninger, CDC.

20　Kevin Toner: *Toner v. Lederle Laboratories*, 112 Idaho 328.

20　Lederle drops out: M. H. Smith, "National Childhood Vaccine Injury Compensation Act," *Pediatrics* 82 (1988): 264–269.

21　Companies stop making other vaccines: Ibid.

21 **Congress passes National Childhood Vaccine Injury Act:** Allen, *Vaccine*, 286.

21 **Reagan signs bill into law:** Ibid., 287.

21 **Vaccine Injury Compensation Program:** E. W. Kitch, G. Evans, and R. Gopin, "U.S. Law," in *Vaccines*, 3rd ed., eds. S. A. Plotkin and W. A. Orenstein (Philadelphia: W. B. Saunders, 1999).

21 **Edward Brandt:** Quoted in Allen, *Vaccine*, 281.

22 **AMA dissatisfied:** Ibid., 282.

22 **Henry Waxman:** Quoted in Colgrove, *State of Immunity*, 215.

22 **Decline in lawsuits:** E. W. Kitch, G. Evans, and R. Gopin, "U.S. Law," in *Vaccines*, 3rd ed., eds. S. A. Plotkin and W. A. Orenstein (Philadelphia: W. B. Saunders, 1999).

3. *A Crude Brew*

25 **Gordon Stewart regarding crude brew:** *DPT: Vaccine Roulette*, WRC-TV, Washington, D.C., April 19, 1982.

25 **Kendrick and Eldering vaccine:** P. Kendrick, G. Eldering, and A. Borowski, "A Study in Active Immunization Against Pertussis," *American Journal of Hygiene* 29 (1939): 133–153.

26 **Pertussis proteins:** K. M. Edwards and M. D. Decker, "Pertussis Vaccines," in *Vaccines*, 5th ed., eds. S. A. Plotkin, W. A. Orenstein, and P. A. Offit (London: Elsevier/Saunders, 2008).

26 **Immunological challenges in vaccines:** P. A. Offit, J. Quarles, M. A. Gerber, et al., "Addressing Parents' Concerns: Do Multiple Vaccines Overwhelm or Weaken the Infant's Immune System?" *Pediatrics* 109 (2002): 124–129.

27 **Larry Baraff regarding reason for study:** *DPT: Vaccine Roulette*, WRC-TV, Washington, D.C., April 19, 1982.

27 **Baraff study:** C. L. Cody, L. J. Baraff, J. D. Cherry, et al., "Nature and Rates of Adverse Reactions Associated with DTP and DT Immunizations in Infants and Children," *Pediatrics* 68 (1981): 650–660.

28 **Baraff regarding safer vaccine:** *DPT: Vaccine Roulette*, WRC-TV, Washington, DC, April 19, 1982.

29 **Geier paper:** M. R. Geier, H. Stanboro, and C. R. Merril, "Endotoxin in Commercial Vaccine," *Applied Environmental Microbiology* 36 (1978): 445–449.

29 **Endotoxin:** T. Gutsmann, A. B. Schromm, and K. Brandenburg, "The Physicochemistry of Endotoxins in Relation to Bioactivity," *International Journal of Medical Microbiology* 297 (2007): 341–352; S. M. Opal, "The Host Response to Endotoxin, Antilipopolysaccharide Strategies, and the Management of Severe Sepsis," *International Journal of Medical Microbiology* 297 (2007): 365–377; S. F. Lowry, "Human Endotoxemia: A Model for Mechanistic Insight and Therapeutic Targeting," *Shock* 24 (2005): 94–100.

29 **Medical Research Council study:** S. Bedson et al., "Vaccination Against Whooping Cough: Report to the Medical Research Council," *British Medical Journal* 2 (1956): 454–462.

29 **EEG study:** B. Hellström, "Electroencephalographic Studies in Triple-Immunized Infants," *British Medical Journal* 2 (1962): 1089–1091.

29 **North West Thames Region study:** T. M. Pollack and J. Morris, "A 7-Year Survey of Disorders Attributed to Vaccination in North West Thames Region," *The Lancet* 1 (1983): 753–757.

30 **Neuropathological study:** J.A.N. Corsellis, I. Janoti, and A. K. Marshall, "Immunization Against Whooping Cough: A Neuropathological Review," *Neuropathology and Applied Neurobiology* 9 (1983): 261–270.

30 **Denmark study:** W. D. Shields, C. Nielson, D. Buch, et al., "Relationship of Pertussis Immunization to the Onset of Neurologic Disorders: A Retrospective Epidemiologic Study," *Journal of Pediatrics* 113 (1988): 801–805.

30 **Group Health Cooperative study:** A. M. Walker, H. Jick, D. R. Perera, et al., "Neurologic Events Following Diphtheria-Tetanus-Pertussis Immunization," *Pediatrics* 81 (1988): 345–349.

30 **Griffin study:** M. R. Griffin, W. A. Ray, E. A. Mortimer, et al., "Risk of Seizures and Encephalopathy After Immunization with the Diphtheria-Tetanus-Pertussis Vaccine," *Journal of the American Medical Association* 263 (1990): 1641–1645.

30 **British Pediatric Association and Canadian National Advisory Committee on Immunization:** J. D. Cherry, "Pertussis Vaccine Encephalopathy: It Is Time to Recognize It as the Myth That It Is," *Journal of the American Medical Association* 263 (1990): 1679–1680; Canadian National Advisory Committee on Immunization, Minister of National Health and Welfare, National Advisory Committee on Immunization, *Canadian Immunization Guide*, 3rd ed. (Ottawa, Canada: Minister of National Health and Welfare, Health Protection Branch, 1989), 78–83.

31 **Copenhagen study:** T. Madsen, "Vaccination Against Whooping Cough," *Journal of the American Medical Association* 101 (1933): 187–188.

31 **Golden review:** G. S. Golden, "Pertussis Vaccine and Injury to the Brain," *Journal of Pediatrics* 116 (1990): 854–861. Italics added for emphasis.

31 **James Cherry and myth:** J. D. Cherry, "Pertussis Vaccine Encephalopathy: It Is Time to Recognize It as the Myth That It Is," *Journal of the American Medical Association* 263 (1990): 1679–1680.

31 **Institute of Medicine report:** Institute of Medicine, *Adverse Effects of Pertussis and Rubella Vaccines: A Report of the Committee to Review the Adverse Consequences of Pertussis and Rubella Vaccines* (Washington D.C.: National Academies Press, 1991).

31 **Child Neurology Society report:** Ad Hoc Committee for the Child Neurology Consensus Statement on Pertussis Immunization and the Central Nervous System, "Pertussis Immunization and the Central Nervous System," *Annals of Neurology* 29 (1991): 458–460.

31 **University of Washington/CDC study:** J. L. Gale, P. B. Thapa, S.G.F. Wassilak, et al., "Risk of Serious Acute Neurological Illness After Immunization with Diphtheria-Tetanus-Pertussis Vaccine: A Population-Based Case-Control Study," *Journal of the American Medical Association* 271 (1994): 37–41.

31 **Revision of VICP table:** G. Evans, E. M. Levine, and E. H. Saindon, "Legal Issues," in *Vaccines*, 5th ed., eds. S. A. Plotkin, W. A. Orenstein, and P. A. Offit (London: Elsevier/Saunders, 2008); 60 Fed. Reg. 7678, 7691 (February 8, 1995).

32 **Computerized records study:** W. E. Barlow, R. L. Davis, J. W. Glasser, et al. for the Centers for Disease Control and Prevention Vaccine Safety DataLink Working Group, "The Risk of Seizures After Receipt of Whole-Cell Pertussis or Measles, Mumps, and Rubella Vaccine," *New England Journal of Medicine* 345 (2001): 656–661.

32 **Febrile seizures:** M. D. Blumstein and M. J. Friedman, "Childhood Seizures," *Emergency Medicine Clinics of North America* 25 (2007): 1061–1086.

32 **Miller study:** D. L. Miller, E. M. Ross, R. Alderslade, et al., "Pertussis Immunization and Serious Acute Neurological Illness in Children," *British Medical Journal* 282 (1981): 1595–1599.

33 **Bonthrone trial:** "Parents Fail in Whooping Cough Vaccination Test Case," *The Times* (London), August 31, 1985.

33 **Johnnie Kinnear's symptoms:** B. Deer, "The Vanishing Victims: Can Whooping Cough Jabs Cause Brain Damage in Children?" *Sunday Times Magazine* (London), November 1, 1998.

34 **Kinnear lawsuit:** "The Law Tries to Decide Whether Whooping Cough Vaccine Causes Brain Damage: Professor Gordon Stewart Gives Evidence," *British Medical Journal* 292 (1986): 1264–1266.

34 **Judge Stuart-Smith:** "Other Major Cases That Brought Him in the Spotlight: Hillsborough," http://www.parliament.the-stationery-office.co.uk/pa/cm199798/cmhansrd/vo980218/debtext/80218-21.htm.

34 **Stuart-Smith and the Hillsborough soccer tragedy:** D. Conn, "Hillsborough: How Stories of Disaster Police Were Altered," *The Guardian*, April 13, 2009.

34 **Stewart at Kinnear trial:** "The Law Tries to Decide Whether Whooping Cough Vaccine Causes Brain Damage: Professor Gordon Stewart Gives Evidence," *British Medical Journal* 292 (1986): 1264–1266.

34 **Kinnear hospital records:** B. Deer, "The Vanishing Victims: Can Whooping Cough Jabs Cause Brain Damage in Children?" *Sunday Times Magazine* (London), November 1, 1998.

34 **Stuart-Smith regarding mother's testimony in Kinnear trial:** Stuart-Smith, Lord Justice, Judgment 30 March 1988, *Loveday v. Renton and the Wellcome Foundation Ltd.*, Chilton Vint & Co, 24 Chauncery Lane, London, WC2.

35 Stewart misrepresents Miller study: "The Law Tries to Decide Whether Whooping Cough Vaccine Causes Brain Damage: Professor Gordon Stewart Gives Evidence," *British Medical Journal* 292 (1986): 1264–1266.

35 Levine and Wenk quote: Ibid.

35 Machin-Stewart interchange: B. Deer, "The Vanishing Victims: Can Whooping Cough Jabs Cause Brain Damage in Children?" *Sunday Times Magazine* (London), November 1, 1998.

35 "An evidential liability": Stuart-Smith, Lord Justice, Judgment 30 March 1988, *Loveday v. Renton and the Wellcome Foundation Ltd.*, Chilton Vint & Co, 24 Chauncery Lane, London, WC2.

35 Loveday trial: Stuart-Smith, Lord Justice, Judgment 30 March 1988, *Loveday v. Renton and the Wellcome Foundation Ltd.*, Chilton Vint & Co, 24 Chauncery Lane, London, WC2.

35 Susan Loveday's story: C. Dyer, "Whooping Cough Vaccine on Trial Again," *British Medical Journal* 295 (1987): 1053–1054.

36 Loveday's story unravels: B. Deer, "The Vanishing Victims: Can Whooping Cough Jabs Cause Brain Damage in Children?" *Sunday Times Magazine* (London), November 1, 1998.

36 Loveday trial: C. Bowie, "Lessons from the Pertussis Vaccine Court Trial," *The Lancet* 335 (1990): 397–399; all quotes from the Loveday trial were obtained from Stuart-Smith, Lord Justice, Judgment 30 March 1988, *Loveday v. Renton and the Wellcome Foundation Ltd.*, Chilton Vint & Co, 24 Chauncery Lane, London, WC2.

39 David Miller's study falls apart: J. D. Cherry, "The Epidemiology of Pertussis and Pertussis Immunization in the United Kingdom and the United States: A Comparative Study," *Current Problems in Pediatrics* 14 (1984): 7–77; J. D. Cherry, "Recurrent Seizures After Diphtheria, Tetanus, and Pertussis Immunization," *American Journal of Diseases in Children* 138 (1984): 904–907; J. D. Cherry, P. A. Brunell, G. S. Golden, and D. T. Karzon, "Report of the Task Force on Pertussis and Pertussis Immunization—1988," *Pediatrics* 81 (1988): 933–984; J.B.P. Stephenson, "A Neurologist Looks at Neurological Disease Temporally Related to DTP Immunization," *Tokai Journal of Experimental and Clinical Medicine* 13 (1988): 157–164.

39 Salisbury on Stuart-Smith verdict: Author interview with David Salisbury, November 30, 2009.

39 Canadian verdict: *Rothwell v. Connaught Laboratories*, Supreme Court of Ontario, June 17, 1988.

40 Seizures in childhood: M. D. Blumstein and M. J. Friedman, "Childhood Seizures," *Emergency Medicine Clinics of North America* 25 (2007): 1061–1086.

40 Seizure types and genetics: R. Nabbout and O. Dulac, "Epileptic Syndromes in Infancy and Childhood," *Current Opinion in Neurology* 21

(2008): 161–166; L. Deprez, A. Jansen, and P. De Jonghe, "Genetics of Epilepsy Syndromes Starting in the First Year of Life," *Neurology* 72 (2009): 273–281.

41 **Berkovic on Dravet's Syndrome:** Author interview with Samuel Berkovic, November 27, 2009.

42 **Berkovic study:** S. F. Berkovic, L. Harkin, J. M. McMahon, et al., "De-Novo Mutations of the Sodium Channel Gene *SCN1A* in Alleged Vaccine Encephalopathy: A Retrospective Study," *Lancet Neurology* 5 (2006): 488–492.

42 **Berkovic regarding reaction to study:** Author interview with Samuel Berkovic, November 27, 2009.

43 **Marge Grant at Hawkins hearing:** Hearings before the Subcommittee on Investigations and General Oversight of the Committee on Labor and Human Resources, U.S. Senate, Ninety-Seventh Congress, Second Session to Examine Adverse Drug Reactions from Immunization, Federal Efforts in Preventive Medicine, and Characteristics of Certain Diseases, May 7, 1982 (emphasis in original text). All Grant quotes below come from this source.

44 **Berkovic study:** S. F. Berkovic, L. Harkin, J. M. McMahon, et al., "De-Novo Mutations of the Sodium Channel Gene *SCN1A* in Alleged Vaccine Encephalopathy: A Retrospective Study," *Lancet Neurology* 5 (2006): 488–492.

44 **Shorvon and Berg editorial:** S. Shorvon and A. Berg, "Pertussis Vaccination and Epilepsy—An Erratic History, New Research and the Mismatch Between Science and Social Policy," *Epilepsia* 49 (2008): 219–225.

44 **Media coverage of Berkovic paper:** Author interview with Samuel Berkovic, November 27, 2009.

4. Roulette Redux

45 **Lea Thompson's statement of purpose:** *DPT: Vaccine Roulette*, WRC-TV, Washington, D.C., April 19, 1982. All Thompson quotes in the section are from this program; italics added for emphasis.

45 **Baraff study:** C. L. Cody, L. J. Baraff, J. D. Cherry, et al., "Nature and Rates of Adverse Reactions Associated with DTP and DT Immunizations in Infants and Children," *Pediatrics* 68 (1981): 650–660.

46 **Pertussis vaccine and SIDS:** R. H. Bernier, J. A. Frank, T. Dondero, and P. Turner, "Diphtheria-Tetanus Toxoids-Pertussis Vaccination and Sudden Infant Deaths in Tennessee," *Journal of Pediatrics* 101 (1982): 419–421.

46 **Infantile spasms study:** J. C. Melchior, "Infantile Spasms and Early Immunization Against Whooping Cough: Danish Survey from 1970 to 1975," *Archives of Disease in Childhood* 52 (1977): 134–137.

46 **Studies confirming Denmark study:** M. H. Bellman, E. M. Ross, and D. L. Miller, "Infantile Spasms and Pertussis Immunization," *The Lancet*

1 (1983): 1031–1033; C. T. Lombroso, "A Prospective Study of Infantile Spasms: Clinical and Therapeutic Correlations," *Epilepsia* 24 (1983): 135–158.

47 **Pertussis in 1982:** K. M. Edwards and M. D. Decker, "Pertussis Vaccines," in *Vaccines*, 5th ed., eds. S. A. Plotkin, W. A. Orenstein, and P. A. Offit (London: Elsevier/Saunders, 2008).

47 **William Foege regarding *DPT: Vaccine Roulette*:** Hearings before the Subcommittee on Investigations and General Oversight of the Committee on Labor and Human Resources, U.S. Senate, Ninety-Seventh Congress, Second Session to Examine Adverse Drug Reactions from Immunization, Federal Efforts in Preventive Medicine, and Characteristics of Certain Diseases, May 7, 1982.

48 **Robert Mendelsohn's medical affiliations:** E. R. González, "TV Report on DTP Galvanizes US Pediatricians," *Journal of the American Medical Association* 248 (1982): 12–22.

48 **Mendelsohn on vaccines and other treatments:** http://whale.to/v/mendelsohn.html.

48 **Mendelsohn on doctors:** http://whale.to/vaccine/quotes20.html.

48 **Mendelsohn on surgeons:** Foreword by Robert S. Mendelsohn, M.D., to Hans Ruesch's *Slaughter of the Innocent* (London: Civitas Publications, 1983).

49 **Mendelsohn and "apple pie":** E. R. González, "TV Report on DTP Galvanizes US Pediatricians," *Journal of the American Medical Association* 248 (1982): 12–22.

49 **Bobby Young's affiliations:** Ibid.

49 **Seizures and brain damage:** M. D. Blumstein and M. J. Friedman, "Childhood Seizures," *Emergency Medicine Clinics of North America* 25 (2007): 1061–1086; J. H. Ellenberg, D. G. Hirtz, and K. B. Nelson, "Do Seizures Cause Intellectual Deterioration?" *New England Journal of Medicine* 314 (1986): 1085–1088.

49 **Thompson on journalistic integrity:** E. R. González, "TV Report on DTP Galvanizes US Pediatricians," *Journal of the American Medical Association* 248 (1982): 12–22.

49 **Edward Mortimer on *Vaccine Roulette*:** Ibid.

50 **Gordon Stewart's affiliations:** Ibid.

50 **Stewart regarding pertussis vaccine efficacy:** G. T. Stewart, "Vaccination Against Whooping Cough: Efficacy Versus Risks," *The Lancet* 1 (1977): 234–237.

50 **Pertussis control by vaccine:** E. J. Gangarosa, A. M. Galazka, C. R. Wolfe, et al., "Impact of Anti-Vaccine Movements on Pertussis Control: The Untold Story," *The Lancet* 351 (1998): 356–361.

50 **Stewart regarding upcoming pertussis outbreak:** "Vaccine Call Is Attacked," *Sunday Times* (London), June 26, 1977.

50 **Pertussis epidemic in England:** J. D. Cherry, "The Epidemiology of Pertussis and Pertussis Immunization in the United Kingdom and the United

States: A Comparative Study," *Current Problems in Pediatrics* 14 (1984): 7–77.

51 **Luc Montagnier discovers HIV:** F. Barré-Sinoussi, J. C. Chermann, F. Rey, et al., "Isolation of a T-Lymphotropic Retrovirus from a Patient at Risk for Acquired Immune Deficiency Syndrome (AIDS)," *Science* 220 (1983): 868–871.

51 **Stewart and sperm:** G. T. Stewart, "The Epidemiology and Transmission of AIDS: A Hypothesis Linking Behavioural and Biological Determinants to Time, Person, and Place," *Genetica* 95 (1995): 173–193.

51 **Stewart and yeast:** E. Papadopulos-Eleopulos, V. G. Turner, J. M. Papadimitriou, et al., "HIV Antibodies: Further Questions and a Plea for Clarification," *Current Medical Research and Opinion* 13 (1997): 627–634.

51 **Stewart and rock stars:** G. Stewart, "AIDS, the Myths and Martyrdom," *Daily Mail* (London), April 5, 1993.

51 **Stewart and behavior of victim:** Ibid.

51 **Stewart and AZT:** G. T. Stewart, "The Epidemiology and Transmission of AIDS: A Hypothesis Linking Behavioural and Biological Determinants to Time, Person, and Place," *Genetica* 95 (1995): 173–193.

53 **John Stossel interviews Allen McDowell:** "Scared Stiff: Worry in America," *ABC News: 20/20*, February 23, 2007.

53 **Anthony Colantoni:** C. McHugh, "Ex-Lawyer Accused of $1.4 Million Fraud," *Chicago Daily Law Bulletin*, November 15, 1993; J. F. Rooney, "Converted Children's Compensation: Ex-Lawyer Admits," *Chicago Daily Law Bulletin*, November 22, 1993.

54 **Thompson in 1997:** First International Conference on Vaccination, National Vaccine Information Center, 1997.

54 **Yellow fever vaccine and hepatitis:** J. P. Fox, C. Manso, H. A. Penna, and M. Para, "Observations on the Occurrence of Icterus in Brazil Following Vaccination Against Yellow Fever," *American Journal of Hygiene* 36 (1942): 68–116; M. V. Hargett and H. W. Burruss, "Aqueous-Based Yellow Fever Vaccine," *Public Health Reports* 58 (1943): 505–512; W. A. Sawyer, K. F. Meyer, M. D. Eaton, et al., "Jaundice in Army Personnel in the Western Region of the United States and Its Relation to Vaccination Against Yellow Fever," *American Journal of Hygiene* 40 (1944): 35–107; L. B. Seeff, G. W. Beebe, J. H. Hoofnagle, et al., "A Serologic Follow-Up of the 1942 Epidemic of Post-Vaccination Hepatitis in the United States Army," *New England Journal of Medicine* 316 (1987): 965–970.

55 **War ended:** G. Williams, *Virus Hunters* (New York: Alfred Knopf, 1960), 315.

55 **Cutter Laboratories tragedy:** Offit, *Cutter*.

55 **Lubeck disaster:** F. Luelmo, "BCG Vaccination," *American Review of Respiratory Diseases* 125 (1982): 70–72.

5. Make the Angels Weep

57 Praise for *A Shot in the Dark*: P. Holt, "Shedding Light on a Drug Controversy," *San Francisco Chronicle*, February 8, 1985.

58 Sabin's vaccine: Oshinsky, *Polio*.

59 Egg protein allergy: B. Ratner and S. Untracht, "Egg Allergy in Children," *American Journal of Diseases of Children* 83 (1952): 309–316.

59 Gelatin allergy: M. Sakaguchi, T. Nakayama, and S. Inouye, "Food Allergy to Gelatin in Children with Systemic Immediate-Type Reactions, Including Anaphylaxis, to Vaccines," *Journal of Allergy and Clinical Immunology* 98 (1996): 1058–1061; M. Sakaguchi, T. Yamanaka, K. Ikeda, et al., "IgE-Mediated Systemic Reactions to Gelatin Included in the Varicella Vaccine," *Journal of Allergy and Clinical Immunology* 99 (1997): 263–264.

61 Hib vaccine licensed: Centers for Disease Control and Prevention, "Haemophilus B Conjugate Vaccines for Prevention of Haemophilus Influenzae Type B Disease Among Infants and Children Two Months of Age and Older: Recommendations of the ACIP," *Morbidity and Mortality Weekly Report* 40 (1991): 1–7.

61 Barbara Loe Fisher on TV: *World News Tonight with Peter Jennings*, February 16, 1998.

61 Fisher and chronic diseases: H. L. Coulter and B. L. Fisher, *A Shot in the Dark: Why the P in the DPT Vaccination May Be Hazardous to Your Children's Health* (Garden City Park, N.Y.: Avery Publishing, 1991).

61 Classen patents: http://www.vaccines.net/patentin.htm.

61 Classen Hib study: D. C. Classen and J. B. Classen, "The Timing of Pediatric Immunization and the Risk of Insulin-Dependent Diabetes Mellitus," *Infectious Diseases Clinical Practitioner* 6 (1997): 449–454.

61 Classen on TV: *World News Tonight with Peter Jennings*, February 16, 1998.

62 First Hib vaccine-diabetes study in American children: S. B. Black, E. Lewis, H. Shinefield, et al., "Lack of Association Between Receipt of Conjugate Haemophilus Influenzae Type B Vaccine (HbOC) in Infancy and Risk of Type 1 (Juvenile Onset) Diabetes: Long-Term Follow-Up of the HbOC Efficacy Trial Cohort," *Pediatric Infectious Disease Journal* 21 (2002): 568–569.

62 Second Hib vaccine-diabetes study in American children: F. DeStefano, J. P. Mullooly, C. A. Okoro, et al., "Childhood Vaccinations, Vaccination Timing, and Risk of Type 1 Diabetes Mellitus," *Pediatrics* 108 (2001), http://www.pediatrics.org/cgi/content/full/108/6/e112.

62 Follow-up Hib vaccine-diabetes study in Finnish children: Institute for Vaccine Safety Diabetes Workshop Panel, "Childhood Immunizations and Type 1 Diabetes: Summary of an Institute for Vaccine Safety Workshop," *Pediatric Infectious Disease Journal* 18 (1999): 217–222.

62   **Heather Whitestone:** http://bhamwiki.com/w/Heather_Whitestone; http://www.heatherwhitestone.com/site.content/bio.shtml; http://www.heather whitestone.com/site/content/faqs.shtml.

62   **Fisher regarding Whitestone:** S. Evans, "How Safe Are Mandatory Immunizations? Doctors Stress That Dangers of Childhood Diseases Far Outweigh Risks of Shots," *Washington Post*, September 27, 1994.

63   **Ted Williams and Whitestone illness:** Ibid.

63   **Fisher regarding conspiracy:** National Vaccine Information Center Archives, volume 1, number 1, March 1995.

64   **Hepatitis B disease in the United States:** E. E. Mast and J. W. Ward, "Hepatitis B Vaccines," in *Vaccines*, 5th ed., eds. S. A. Plotkin, W. A. Orenstein, and P. A. Offit (London: Elsevier/Saunders, 2008).

65   **Hepatitis B vaccine policy in the United States:** Ibid.

65   **Hepatitis B vaccine discussed on 20/20:** *ABC News 20/20*, January 22, 1999.

67   **Fisher regarding hepatitis B virus infections in children:** S. H. Trudeau, "Is There a Risk to Vaccines?" *Copley News Service*, January 12, 1998.

67   **Hepatitis B disease in children:** G. L. Armstrong, E. E. Mast, M. Wojczynski, and H. S. Margolis, "Childhood Hepatitis B Virus Infections in the United States Before Hepatitis B Immunization," *Pediatrics* 108 (2001): 1123–1128.

67   **Risk of cirrhosis and liver cancer following childhood infection:** E. E. Mast and J. W. Ward, "Hepatitis B Vaccines," in *Vaccines*, 5th ed., eds. S. A. Plotkin, W. A. Orenstein, and P. A. Offit (London: Elsevier/Saunders, 2008).

67   **Risk of chronic disease:** Ibid.

67   **Belkin holding CDC accountable:** "Vaccine Safety Group Endorses Government Action to Eliminate Mercury in Childhood Vaccines and Roll Back Hepatitis B Vaccination for Most Newborn Infants," *PR Newswire*, July 8, 1999.

68   **Belkin regarding junkies and homosexuals:** M. Benjamin, "UPI Investigates: The Vaccine Conflict," *United Press International*, July 20, 2003.

68   **Studies exonerate hepatitis B vaccine as cause of SIDS:** E. A. Mitchell, A. W. Stewart, and M. Clements, "Immunisation and the Sudden Infant Death Syndrome: New Zealand Cot Death Study Group," *Archives of Diseases of Childhood* 73 (1995): 498–501; M. T. Niu, M. E. Salive, and S. S. Ellenberg, "Neonatal Deaths and Hepatitis B Vaccine: The Vaccine Adverse Event Reporting System, 1991–1998," *Archives of Pediatric and Adolescent Medicine* 153 (1999): 1279–1282; E. M. Eriksen, J. A. Perlman, A. Miller, et al., "Lack of Association Between Hepatitis B Birth Immunization and Neonatal Death: A Population-Based Study from the Vaccine Safety DataLink Project," *Pediatric Infectious Disease Journal* 23 (2004): 656–662.

68   **Biological arguments against hepatitis B vaccine as a cause of multiple sclerosis:** P. A. Offit and C. J. Hackett, "Addressing Parents' Concerns:

Do Vaccines Cause Allergic or Autoimmune Diseases?" *Pediatrics* 111 (2003): 653–659.

68 **Studies exonerating hepatitis B vaccine as a cause of multiple sclerosis:** A. Ascherio, S. M. Zhang, M. A. Hernan, et al., "Hepatitis B Vaccination and the Risk of Multiple Sclerosis," *New England Journal of Medicine* 344 (2001): 327–332; C. Confavreux, S. Suissa, P. Saddier, et al., "Vaccinations and the Risk of Relapse in Multiple Sclerosis," *New England Journal of Medicine* 344 (2001): 319–326.

68 **Experience with hepatitis B vaccine:** E. E. Mast and J. W. Ward, "Hepatitis B Vaccines," in *Vaccines*, 5th ed., eds. S. A. Plotkin, W. A. Orenstein, and P. A. Offit (London: Elsevier/Saunders, 2008).

69 **Pneumococcal vaccine study:** S. Black, H. Shinefield, B. Fireman, et al., "Efficacy, Safety, and Immunogenicity of Heptavalent Pneumococcal Conjugate Vaccine in Children: Northern California Kaiser Permanente Study Center Group," *Pediatric Infectious Disease Journal* 19 (2000): 183–186.

69 **TV report on pneumococcal vaccine:** "New Pneumococcal Vaccine for Children," *World News Tonight with Peter Jennings*, September 25, 1998.

70 **Pneumococcal disease:** S. Black, J. Eskola, C. Whitney, and H. Shinefield, "Pneumococcal Conjugate Vaccine and Pneumococcal Common Protein Vaccines," in *Vaccines*, 5th ed., eds. S. A. Plotkin, W. A. Orenstein, and P. A. Offit (London: Elsevier/Saunders, 2008).

71 **Daum and "giant step":** L. Richwine, "FDA Panel Deems New Vaccine Safe: Experts Called It Effective on Resistant Bacteria That Kill a Million Youths a Year," *Philadelphia Inquirer*, November 6, 1999.

71 **Fisher and not enough evidence:** Ibid.

71 **Fisher and "post-marketing experiment":** "FDA Advisers Back Safety and Efficacy of Wyeth's Pneumococcal Vaccine for Children," *Reuters*, November 8, 1999.

71 **Effect of pneumococcal vaccine:** S. Black, J. Eskola, C. Whitney, and H. Shinefield, "Pneumococcal Conjugate Vaccine and Pneumococcal Common Protein Vaccines," in *Vaccines*, 5th ed., eds. S. A. Plotkin, W. A. Orenstein, and P. A. Offit (London: Elsevier/Saunders, 2008); T. Pillshvili, C. Lexau, M. M. Farley, et al., "Sustained Reductions in Invasive Pneumococcal Disease in the Era of Conjugate Vaccine," *Journal of Infectious Diseases* 201 (2010): 32–41.

72 **Rotavirus vaccine and disease:** H. F. Clark, P. A. Offit, R. I. Glass, and R. M. Ward, "Rotavirus Vaccines," in *Vaccines*, 5th ed., eds. S. A. Plotkin, W. A. Orenstein, and P. A. Offit (London: Elsevier/Saunders, 2008).

72 **VAERS reports of intussusception following rotavirus vaccine:** Centers for Disease Control and Prevention, "Intussusception Among Recipients of Rotavirus Vaccine—United States, 1998–1999," *Morbidity and Mortality Weekly Report* 48 (1999): 577–581.

72 **Rotavirus vaccine causes intussusception:** T. V. Murphy, P. M. Garguillo, M. S. Massoudi, et al., "Intussusception Among Infants Given an Oral

Rotavirus Vaccine," *New England Journal of Medicine* 344 (2001): 564–572; P. Kramarz, E. K. France, F. DeStefano, et al., "Population-Based Study of Rotavirus Vaccination and Intussusception," *Pediatric Infectious Disease Journal* 20 (2001): 410–416.

73 **HPV disease:** J. T. Schiller, I. H. Frazer, and D. R. Lowry, "Human Papillomavirus Vaccines," in *Vaccines*, 5th ed., eds. S. A. Plotkin, W. A. Orenstein, and P. A. Offit (London: Elsevier/Saunders, 2008).

73 **HPV vaccine:** Ibid.

73 **Fisher regarding flawed science:** "Merck's Gardasil Vaccine Not Proven Safe for Little Girls; National Vaccine Information Center Criticizes FDA for Fast-Tracking Licensure," *PR Newswire*, June 27, 2006.

74 **Fisher and "death by Vioxx":** "HPV Vaccine Now, HIV Vaccine Next," posted by Barbara Loe Fisher, August 16, 2006, http://www. vaccine awakening.blogspot.com.

74 **Fisher and "the slut shot":** "The Slut Shot," posted by Barbara Loe Fisher, August 17, 2006, http://www.vaccineawakening.blogspot.com.

74 **Fisher and "the 'cheaters' vaccine":** "The 'Cheaters' Vaccine: HPV," posted by Barbara Loe Fisher, October 3, 2006, http://www.vaccine awakening.blogspot.com.

74 **Epidemiology of HPV in American women:** H. Richardson, G. Kelsall, P. Tellier, et al., "Natural History of Type-Specific Human Papillomavirus Infections in Female University Students," *Cancer Epidemiology Biomarkers and Prevention* 12 (2003): 485–490; R. L Winer, S. K. Lee, J. P. Hughes, et al., "Genital Human Papillomavirus Infection: Incidence and Risk Factors in a Cohort of Female University Students," *American Journal of Epidemiology* 157 (2003): 218–226.

74 **Fisher on TV:** *CBS News Sunday Morning*, April 1, 2007.

74 **CDC evaluates HPV vaccine safety:** http://www.cdc.gov/vaccinesafety .vaers.gardasil.htm.

74 **HPV vaccine and GBS:** Ibid.; S. L. Block, D. R. Brown, A. Chatterjee, et al., "Clinical Trial and Post-Licensure Safety Profile of a Prophylactic Human Papillomavirus (Types 6, 11, 16, and 18) L1 Virus-Like Particle Vaccine," *Pediatric Infectious Disease Journal* 29 (2010): 95–101.

75 **Fisher discounts coincidence:** "Gardasil Vaccine: The Damage Continues," posted by Barbara Loe Fisher, August 15, 2008, http://www .vaccineawakening. blogspot.com.

75 **Fisher and HPV vaccine causing cancer:** "Vaccine Safety Group Releases Gardasil Reaction Reports," *PR Newswire*, February 21, 2007.

75 **Fisher and chickenpox vaccine:** S. H. Trudeau, "Is There a Risk to Vaccines?" *Copley News Service*, January 12, 1998.

75 **Chickenpox:** A. A. Gershon, M. Takahashi, and J. F. Seward, "Varicella Vaccine," in *Vaccines*, 5th ed., eds. S. A. Plotkin, W. A. Orenstein, and P. A. Offit (London: Elsevier/Saunders, 2008).

75 **Fisher and shifting burden of chickenpox by vaccine:** S. H. Trudeau, "Is There a Risk to Vaccines?" *Copley News Service*, January 12, 1998.

75 Chickenpox vaccine efficacy: A. A. Gershon, M. Takahashi, and J. F. Seward, "Varicella Vaccine," in *Vaccines*, 5th ed., eds. S. A. Plotkin, W. A. Orenstein, and P. A. Offit (London: Elsevier/Saunders, 2008).

76 Fisher and herd immunity: H. Collins, "Life Giver or Life Taker: A Debate on the Value of Vaccines, Special Report, Immunizations: A Public Health Staple Comes Under Siege," *Philadelphia Inquirer*, October 3, 1999.

76 Measles outbreak in the Netherlands: S. von den Hof, M.A.E. Conynvan Spaendonck, and J. E. van Steenbergen, "Measles Epidemic in the Netherlands: 1999–2000," *Journal of Infectious Diseases* 186 (2002): 1483–1486.

76 Fisher and unrealistic fear of disease: K. G. Goff, "Measles Makes Unwelcome Return," *Washington Times*, May 6, 2009.

76 Measles: P. M. Strebel, M. J. Papania, G. H. Dayan, and N. A. Halsey, "Measles Vaccine," in *Vaccines*, 5th ed., eds. S. A. Plotkin, W. A. Orenstein, and P. A. Offit (London: Elsevier/Saunders, 2008).

76 Fisher on virtue of natural influenza infection: "Using Religion to Promote Flu Vaccine," posted by Barbara Loe Fisher, January 20, 2007, http://www.vaccineawakening.blogspot.com.

76 Deaths from pandemic influenza, 2009: www.cdc.gov/h1n1flu/update .htm.

77 Berkovic regarding guilt: Author interview with Samuel Berkovic, November 27, 2009.

77 John Salamone's background: National Italian-American Foundation: John Salamone, http://www.niaf.org/about/board_officers.asp.

78 Salamone on son: "Injected Polio Vaccine Winning Support," *Baltimore Sun*, March 29, 1999.

78 Salamone and drunken sailor: Ibid.

78 Salamone on son's diagnosis: Author interview with John Salamone, December 4, 2009. All other quotes from John Salamone in this chapter were obtained from this interview.

81 Fisher quotes Shakespeare: H. L. Coulter and B. L. Fisher, *A Shot in the Dark: Why the P in the DPT Vaccination May Be Hazardous to Your Children's Health* (Garden City Park, N.Y.: Avery Publishing, 1991).

81 Fisher accuses health officials of deception: "Babies Die After MMR," posted by Barbara Loe Fisher, June 19, 2006, http://www.vaccine awakening.blogspot.com.

82 Fisher accuses experts of not caring: "Real Life Experience of a Vaccine Reaction," posted by Barbara Loe Fisher, May 1, 2006, http://www .vaccineawakening.blogspot.com.

82 Fisher and Nazi Germany: "Doctors Want Power to Kill Disabled Babies," posted by Barbara Loe Fisher, November 5, 2006, http://www .vaccineawakening.blogspot.com.

83 Fisher and pencil pushers: "Pencil Pushers Deny Vaccine/Optic Neuritis Link," posted by Barbara Loe Fisher, June 19, 2006, http://www .vaccineawakening.blogspot.com.

83 **Fisher and "bloodbath":** Wingspread meeting headed by Dr. Roger
    Bernier, July 2002; personal communication from Dr. Deborah Wexler,
    April 9, 2010.

6. Justice

86 **VICP:** G. Evans, E. M. Levine, and E. H. Saindon, "Legal Issues," in *Vaccines*, 5th ed., eds. S. A. Plotkin, W. A. Orenstein, and P. A. Offit (London: Elsevier/Saunders, 2008).

86 **Rett Syndrome:** Ibid.

86 **SIDS and vaccines:** Ibid.

86 **DTP vaccine and attention deficit disorder:** Allen, *Vaccine*, 287.

86 **Lawyer admission:** Ibid.

86 **Dog gets stupider:** Ibid.

86 **VICP change:** G. Evans, E. M. Levine, and E. H. Saindon, "Legal Issues," in *Vaccines*, 5th ed., eds. S. A. Plotkin, W. A. Orenstein, and P. A. Offit (London: Elsevier/Saunders, 2008).

86 **Fisher regarding VICP change:** Barbara Loe Fisher testimony before the U.S. House Government Reform Committee, "Vaccines: Finding a Balance Between Public Safety and Personal Choice," August 3, 1999; and testimony before the House Subcommittee on Criminal Justice Drug Policy and Human Resources, House Government Reform Civil Service Compensation for Vaccine Injuries, September 28, 1999.

86 **Margaret Althen:** *Althen v. Department of Health and Human Services*, 418 F.3d 1274 (Fed. Cir. 2005).

87 **Vaccine and demyelinating diseases:** J. Tuttle, R. T. Chen, H. Rantala, et al., "The Risk of Guillain-Barré Syndrome After Tetanus-Toxoid-Containing Vaccines in Adults and Children in the United States," *American Journal of Public Health* 87 (1997): 2045–2048; T. Verstraeten, R. Davis, F. DeStefano, and the Vaccine Safety DataLink Team, "Decreased Risk of Demyelinating Disease Following Tetanus Immunization [abstract]," *American Journal of Epidemiology* 155 (2002): S52; C. Confavreux, S. Suissa, P. Saddier, et al., "Vaccinations and the Risk of Relapse in Multiple Sclerosis," *New England Journal of Medicine* 344 (2001): 319–326.

87 **Rose Capizzano:** *Capizzano v. Department of Health and Human Services*, 440 F.3d 1317 (Fed. Cir. 2006) (quotations regarding the case are from this document); G. Evans, E. M. Levine, and E. H. Saindon, "Legal Issues," in *Vaccines*, 5th ed., eds. S. A. Plotkin, W. A. Orenstein, and P. A. Offit (London: Elsevier/Saunders, 2008).

87 **Hepatitis B vaccine and rheumatoid arthritis:** E. E. Mast and J. W. Ward, "Hepatitis B Vaccines," in *Vaccines*, 5th ed., eds. S. A. Plotkin, W. A. Orenstein, and P. A. Offit (London: Elsevier/Saunders, 2008).

87 **Hepatitis B vaccine and paralysis:** *Stevens v. Secretary of the Department of Health and Human Services*, March 30, 2001.

87 **Hib vaccine and paralysis:** *Camerlin v. Secretary of the Department of Health and Human Services*, October 29, 2003.

87 **MMR and epilepsy:** *Cusati v. Secretary of the Department of Health and Human Services*, September 22, 2005.

87 **MMR and fibromyalgia:** *Zatuchni v. Secretary of the Department of Health and Human Services*, February 12, 2008.

87 **Hepatitis B vaccine and Guillain-Barré Syndrome:** *Peugh v. Secretary of the Department of Health and Human Services*, April 21, 2006.

87 **Rubella vaccine and chronic arthritis:** P. E. Slater, T. Ben Zvi, A. Fogel, et al., "Absence of an Association Between Rubella Vaccination and Arthritis in Underimmune Post-Partum Women," *Vaccine* 13 (1995): 1529–1532; P. Ray, S. Black, H. Shinefield, et al., "Risk of Chronic Arthropathy Among Women After Rubella Vaccination," *Journal of the American Medical Association* 278 (1997): 551–556.

87 **Evidence exonerating vaccines as cause of paralysis, epilepsy, and arthritis:** Ibid.; E. E. Mast and J. W. Ward, "Hepatitis B Vaccines," in *Vaccines*, 5th ed., eds. S. A. Plotkin, W. A. Orenstein, and P. A. Offit (London: Elsevier/Saunders, 2008).

88 **Werderitsch decision:** *Werderitsch v. Secretary of the Department of Health and Human Services*, May 26, 2006.

88 **Epidemiology of multiple sclerosis and hepatitis B virus infections:** E. E. Mast and J. W. Ward, "Hepatitis B Vaccines," in *Vaccines*, 5th ed., eds. S. A. Plotkin, W. A. Orenstein, and P. A. Offit (London: Elsevier/Saunders, 2008).

88 **Hepatitis B vaccine not a cause of multiple sclerosis:** A. Ascherio, S. M. Zhang, M. A. Hernan, et al., "Hepatitis B Vaccination and the Risk of Multiple Sclerosis," *New England Journal of Medicine* 344 (2001): 327–332; C. Confavreux, S. Suissa, P. Saddier, et al., "Vaccinations and the Risk of Relapse in Multiple Sclerosis," *New England Journal of Medicine* 344 (2001): 319–326.

89 **Rorke-Adams's background and training:** L. B. Rorke-Adams, "Lucy Balian Rorke-Adams, MD: An Autobiography," *Journal of Child Neurology* 23 (2008): 674–682.

90 **Rorke-Adams on her experiences in the Vaccine Injury Compensation Program:** Author interview with Lucy Rorke-Adams, April 28, 2009.

90 **Rorke-Adams on John Shane:** Ibid.; additional Rorke-Adams quotes are from this interview.

90 **Rorke-Adams's expertise on embryonic neuronal cells:** An extensive publication list can be found in L. B. Rorke, "Embryonal Tumors of the Central Nervous System," in *Principles and Practice of Neuropathology*, 2nd ed., eds. J. S. Nelson, H. Mena, J. E. Parisi, and S. S. Schochet (New York: Oxford University Press, 2003).

91 **Rorke-Adams's monograph on myelination:** L. B. Rorke and H. E. Riggs, *Myelination of the Brain in the Newborn* (Philadelphia: J. B. Lippincott, 1969).

91 **Karoly and Shane indictment:** *United States of America v. John P. Karoly, Jr., John J. Shane, and John P. Karoly, III,* filed in U.S. District Court for the Eastern District of Pennsylvania, September 25, 2008.

91 **Laurie Magid on Karoly-Shane indictment:** Press release, U.S. Department of Justice, U.S. Attorney, Eastern District of Pennsylvania, "Allentown Attorney and Two Others Indicted for Fraud Involving Couple Killed in Plane Crash," September 25, 2008.

91 **Karoly and tax evasion:** "Allentown Attorney Charged with Defrauding Charity and Church in $500,000 Scheme," Department of Justice Press Release, United States Attorney's Office, Eastern District of Pennsylvania, March 12, 2009; B. Theodore, "Allentown Lawyer John Karoly Jr. Pleads Guilty to Tax Evasion, Could Face Prison Time," LehighValleyLive.com, July 6, 2009; M. Birkbeck, "'We Trusted John Karoly,' Says Former Local Pastor," *The Morning Call,* September 16, 2009; *United States of America v. John P. Karoly, Jr.,* "Government's Guilty Plea Memorandum," United States District Court for the Eastern District of Pennsylvania, Criminal No. 08-592-01.

91 **Fisher on autism:** H. L. Coulter and B. L. Fisher, *A Shot in the Dark: Why the P in the DPT Vaccination May Be Hazardous to Your Children's Health* (Garden City, N.Y.: Avery Publishing, 1991).

92 **Wakefield paper:** A. J. Wakefield, S. H. Murch, A. Anthony, et al., "Ileal-Lymphoid-Nodular Hyperplasia, Non-Specific Colitis, and Pervasive Developmental Disorder in Children," *The Lancet* 351 (1998): 637–641.

92 **Salisbury on fears of pertussis and MMR vaccine:** Author interview with David Salisbury, November 30, 2009.

92 **Measles outbreaks:** N. Gould, "The Town Divided by a Deadly Disease," *Belfast Telegraph,* November 14, 2004; "Fall in MMR Vaccine Coverage Reported as Further Evidence of Vaccine Safety Is Published," *CDR Weekly,* June 25, 1999; B. Lavery, "As Vaccination Rates Decline in Ireland, Cases of Measles Soar," *New York Times,* February 8, 2003; T. Peterkin, "Alert over 60 Percent Rise in Measles," *London Daily Telegraph,* May 12, 2003; N. Begg, M. Ramsey, J. White, and Z. Bozoky, "Media Dents Confidence in MMR Vaccine," *British Medical Journal* 316 (1998): 561; B. Deer, "Schoolboy, 13, Dies as Measles Makes a Comeback," *Sunday Times* (London), April 2, 2006; K. Mansey, "MMR Link to Mumps Cases," *Daily Post,* January 16, 2006; S. Boseley, "MMR Vaccinations Fall to New Low," *The Guardian,* September 24, 2004; E. K. Mulholland, "Measles in the United States, 2006," *New England Journal of Medicine* 355 (2006): 440–443; J. McBrien, J. Murphy, D. Gill, et al., "Measles Outbreak in Dublin," *Pediatric Infectious Disease Journal* 22 (2003): 580–584; P. A. Brunell, "More on Measles and the Impact of the Lancet Retraction," *Infectious Diseases in Children,* May 2004; B. Deer, "MMR Scare Doctor Faces List of Charges," *The Times* (London), September 11, 2005; S. Hastings, "Doctor at Sharp End of MMR Controversy," *Yorkshire Post,* June 14, 2006.

92 **Parents in United States refuse MMR vaccine:** M. J. Smith, L. M. Bell, S. E. Ellenberg, and D. M. Rubin, "Media Coverage of the MMR-Autism Controversy and Its Relationship to MMR Immunization Rates in the United States," *Pediatrics* 121 (2008): e836–e843.

93 **Measles virus and intestines of autistic children:** M. Hornig, T. Briese, T. Buie, et al., "Lack of Association Between Measles Virus Vaccine and Autism with Enteropathy: A Case-Control Study," *PLoS ONE* 3 (2008): e3140.

93 **MMR vaccine does not cause intestinal inflammation:** H. Peltola, A. Patja, P. Leinikki, et al., "No Evidence for Measles, Mumps, and Rubella Vaccine-Associated Inflammatory Bowel Disease or Autism in a 14-Year Prospective Study," *The Lancet* 351 (1998): 1327–1328; R. L. Davis, P. Kramarz, B. Kari, et al., "Measles-Mumps-Rubella and Other Measles-Containing Vaccines Do Not Increase the Risk for Inflammatory Bowel Disease: A Case-Control Study from the Vaccine Safety DataLink Project," *Archives of Pediatrics and Adolescent Medicine* 155 (2002): 354–359; B. Taylor, E. Miller, R. Lingam, et al., "Measles, Mumps, and Rubella Vaccination and Bowel Problems or Developmental Regression in Children with Autism: Population Study," *British Medical Journal* 324 (2002): 393–396; E. Fombonne and E. H. Cook, Jr., "MMR and Autistic Enterocolitis: Consistent Epidemiological Failure to Find an Association," *Molecular Psychiatry* 8 (2003): 133–134.

93 **Autism and brain-damaging (encephalopathic) proteins:** K. Wang, H. Zhang, D. Ma, et al., "Common Genetic Variants on 5p14.1 Associate with Autism Spectrum Disorders," *Nature* 459 (2009): 528–533.

93 **Epidemiological studies exonerate MMR vaccine:** B. Taylor, E. Miller, C. P. Farrington, et al., "Autism and Measles, Mumps, and Rubella Vaccine: No Epidemiological Evidence for a Causal Association," *The Lancet* 353 (1999): 2026–2029; F. DeStefano and R. T. Chen, "Negative Association Between MMR and Autism," *The Lancet* 353 (1999): 1986–1987; E. Fombonne, "Are Measles Infections or Measles Immunizations Linked to Autism?" *Journal of Autism and Developmental Disorders* 29 (1999): 349–350; J. A. Kaye, M. Melero-Montes, and H. Jick, "Mumps, Measles, and Rubella Vaccine and the Incidence of Autism Recorded by General Practitioners: A Time Trend Analysis," *British Medical Journal* 322 (2001): 460–463; L. Dales, S. J. Hammer, and N. J. Smith, "Time Trends in Autism and in MMR Immunization Coverage in California," *Journal of the American Medical Association* 285 (2001): 1183–1185; C. P. Farrington, E. Miller, and B. Taylor, "MMR and Autism: Further Evidence Against a Causal Association," *Vaccine* 19 (2001): 3632–3635; E. Fombonne and S. Chakrabarti, "No Evidence for a New Variant of Measles-Mumps-Rubella-Induced Autism," *Pediatrics* 108 (2001), http://www.pediatrics.org/cgi/content/full.108/4/e58; N. A. Halsey and S. L. Hyman, "Measles-Mumps-Rubella Vaccine and Autistic Spectrum Disorder: Report from the New Challenges in Childhood Immunization

Conference Convened in Oak Brook, Illinois, June 12, 2000," *Pediatrics* 107 (2001), www.pediatrics.org/cgi/content/full/107/5/e84; K. M. Madsen, A. Hviid, M. Vestergaard, et al., "A Population-Based Study of Measles, Mumps, and Rubella Vaccination and Autism," *New England Journal of Medicine* 347 (2002): 1477–1482; A. Mäkela, J. P. Nuorti, and H. Peltola, "Neurologic Disorders After Measles-Mumps-Rubella Vaccination," *Pediatrics* 110 (2002): 957–963; P. A. Offit and S. E. Coffin, "Communicating Science to the Public: MMR Vaccine and Autism," *Vaccine* 22 (2003): 1–6; F. DeStefano, T. K. Bhasin, W. W. Thompson, et al., "Age at First Measles-Mumps-Rubella Vaccination in Children with Autism and School-Matched Control Subjects: A Population-Based Study in Metropolitan Atlanta," *Pediatrics* 113 (2004): 259–266; K. Wilson, E. Mills, C. Ross, et al., "Association of Autistic Spectrum Disorder and the Measles, Mumps, and Rubella Vaccine," *Archives of Pediatric and Adolescent Medicine* 157 (2003): 628–634; H. Honda, Y. Shimizu, and M. Rutter, "No Effect of MMR Withdrawal on the Incidence of Autism: A Total Population Study," *Journal of Child Psychiatry and Psychology* 46 (2005): 572–579.

93 **Brian Deer on Andrew Wakefield's conflicts:** B. Deer, "MMR: The Truth Behind the Crisis," *Sunday Times* (London), February 22, 2004.

94 **Wakefield's co-authors withdraw names from paper:** R. Horton, *MMR Science and Fiction: Exploring the Vaccine Crisis* (London: Granta Books, 2004).

94 **Nicholas Chadwick:** Omnibus Autism Proceeding, Federal Claims Court, Washington, D.C., www.uscfc.uscourt.gov/OSM/OSMAutism.htm.

94 **Wakefield wins "Courage in Science Award":** L. Reagan, "Vaccine Conference Exclusive Report," http://www.hpakids.org/holistic-health/articles/187/1/Vaccine-Conference-Exclusive-Report.

94 **Fisher on Madsen study:** Quoted in "Autism: Study Finds No Connection to MMR Vaccine," *American Health Line*, November 7, 2002.

94 **Fisher on Institute of Medicine:** M. Fox, "Study Says Vaccine Not Cause of Autism," *Philadelphia Inquirer*, May 19, 2004.

94 **General Medical Council chastises Wakefield:** N. Triggle, "MMR Doctor 'Broke Research Rules,'" *BBC News*, January 28, 2010; B. Deer, "'Callous, Unethical, and Dishonest': Dr. Andrew Wakefield," TimesONLINE, *Sunday Times* (London), January 31, 2010, http://timesonline.co .uk; Press Association, "MMR Doctor Failed to Act in Interests of Children," *The Guardian*, http://guardian.co.uk/science/2010/jan/28/mmr -doctor-fail.children-gmc/print.

95 **Fisher defends Wakefield after General Medical Council rebuke:** B. L. Fisher, "Vaccines: Doctor Judges and Juries Hanging Their Own," http:// ageofautism.com/2010/01/vaccines-doctor-judges-juries-hanging-their -own.html.

95 ***The Lancet* retracts Wakefield paper:** Editors of *The Lancet*, "Retraction: Ileal-Lymphoid-Nodular Hyperplasia, Non-Specific Colitis, and Perva-

sive Developmental Disorder in Children," *The Lancet*, February 2, 2010.

95 Fisher responds to *The Lancet*'s retraction: T. Miller, "Journal Retracts Study Backing Vaccine-Autism Link," *PBS News Hour*, http://pbs.org/newshour/updates/europe/jan-june10/lancet_0204html.

96 William Sawyer and yellow fever vaccine: W. A. Sawyer, K. F. Meyer, M. D. Eaton, et al., "Jaundice in Army Personnel in the Western Region of the United States and Its Relation to Vaccination Against Yellow Fever," *American Journal of Hygiene* 40 (1944): 35–107.

96 Neil Nathanson and polio vaccine: N. Nathanson and A. D. Langmuir, "The Cutter Incident: Poliomyelitis Following Formaldehyde-Inactivated Poliovirus Vaccination in the United States During the Spring of 1955. I. Background," *American Journal of Hygiene* 78 (1963): 16–28.

96 Trudy Murphy and rotavirus vaccine: T. V. Murphy, P. M. Garguillo, M. S. Massoudi, et al., "Intussusception Among Infants Given an Oral Rotavirus Vaccine," *New England Journal of Medicine* 344 (2001): 564–572.

96 Wakefield resigns: M. A. Roser, "British Doctor Resigns as Head of Austin Autism Center," *Austin American-Statesman*, February 18, 2010.

96 GMC revokes Wakefield's medical license: K. Kelland, "UK Doctor at Heart of Vaccine Row Banned from Practice," *Reuters*, May 24, 2010.

96 AAP and CDC call for removal of thimerosal: P. A. Offit, "Thimerosal and Vaccines: A Cautionary Tale," *New England Journal of Medicine* 357 (2007): 1278–1279.

96 Epidemiological studies of thimerosal and autism: P. Stehr-Green, P. Tull, M. Stellfeld, et al., "Autism and Thimerosal-Containing Vaccines: Lack of Consistent Evidence for an Association," *American Journal of Preventive Medicine* 25 (2005): 101–106; K. M. Madsen, M. B. Lauritsen, C. B. Pedersen, et al., "Thimerosal and the Occurrence of Autism: Negative Ecological Evidence from Danish Population-Based Data," *Pediatrics* 112 (2003): 604–606; A. Hviid, M. Stellfeld, J. Wohlfahrt, and M. Melbye, "Association Between Thimerosal-Containing Vaccine and Autism," *Journal of the American Medical Association* 290 (2003): 1763–1766; J. Heron and J. Golding, "Thimerosal Exposure in Infants and Developmental Disorders: A Prospective Cohort Study in the United Kingdom Does Not Support a Causal Association," *Pediatrics* 114 (2004): 577–583; N. Andrews, E. Miller, A. Grant, et al., "Thimerosal Exposure in Infants and Developmental Disorders: A Retrospective Cohort Study in the United Kingdom Does Not Support a Causal Association," *Pediatrics* 114 (2004): 584–591; E. Fombonne, R. Zakarian, A. Bennett, et al., "Pervasive Developmental Disorders in Montreal, Quebec, Canada: Prevalence and Links with Immunization," *Pediatrics* 118 (2006): 139–150; W. W. Thompson, C. Price, B. Goodson, et al., "Early Thimerosal Exposure and Neuropsychological Outcomes at 7 to 10 Years," *New England Journal of Medicine* 357 (2007): 1281–1292; R.

Schechter and J. Grether, "Continuing Increases in Autism Reported to California's Development Services System," *Archives of General Psychiatry* 65 (2008): 19–24.

96 **Autism rates continued to climb:** R. Schechter and J. Grether, "Continuing Increases in Autism Reported to California's Development Services System," *Archives of General Psychiatry* 65 (2008): 19–24.

97 **Gary Golkiewicz regarding workload:** Speech to the Advisory Commission on Childhood Vaccines, March 6, 2008.

97 **Special master regarding workload:** *Theresa Cedillo and Michael Cedillo v. Secretary of Health and Human Services,* filed February 12, 2009.

98 **Golkiewicz regarding goal of VICP:** Speech to the Advisory Commission on Childhood Vaccines, March 6, 2008.

99 **Special master regarding subjective belief:** *Colten Snyder, Kathryn Snyder, and Joseph Snyder v. Secretary of Health and Human Services,* filed February 12, 2009. Italics added for emphasis.

99 **Special master regarding sentiment:** *Rolf and Angela Hazelhurst and William Yates Hazelhurst v. Secretary of Health and Human Services,* filed February 12, 2009. Italics added for emphasis.

99 **Vote unanimous:** *Theresa Cedillo and Michael Cedillo v. Secretary of Health and Human Services,* filed February 12, 2009; *Rolf and Angela Hazelhurst and William Yates Hazelhurst v. Secretary of Health and Human Services,* filed February 12, 2009; *Colten Snyder, Kathryn Snyder, and Joseph Snyder v. Secretary of Health and Human Services,* filed February 12, 2009.

99 **Special master regarding bad science:** *Colten Snyder, Kathryn Snyder, and Joseph Snyder v. Secretary of Health and Human Services,* filed February 12, 2009. Italics added for emphasis.

99 **Special master and "White Queen":** Ibid.

99 **Special master regarding quality of experts:** Ibid.

100 **Special master on Marcel Kinsbourne's professional testimony:** *Colten Snyder, Kathryn Snyder, and Joseph Snyder v. Secretary of Health and Human Services,* filed February 12, 2009.

100 **Special master on Kinsbourne's statement regarding measles and autism:** Ibid.

100 **Special master on Vera Byers:** Ibid.

101 **Special master on Arthur Krigsman:** *Theresa Cedillo and Michael Cedillo v. Secretary of Health and Human Services,* filed February 12, 2009.

101 **Special master on Jeff Bradstreet's profit:** Ibid.

102 **Verdict on thimerosal cases:** www.uscfc.uscourts.gov.

102 **Special master regarding Krigsman's medical misjudgment:** *Theresa Cedillo and Michael Cedillo v. Secretary of Health and Human Services,* filed February 12, 2009.

102 **Lawyers' compensation:** K. Seidel, "Autism-Vaccine Attorney Bill Tops $2 Million," http://www.neurodiversity.com/weblog/article/180.

103 Special master regarding maturing science: *Colten Snyder, Kathryn Snyder, and Joseph Snyder v. Secretary of Health and Human Services*, filed February 12, 2009.
104 Vaccine shortages: J. Cohen, "U.S. Vaccine Supply Falls Seriously Short," *Science* 295 (2002): 1998–2001; National Vaccine Advisory Committee, "Strengthening the Supply of Routinely Recommended Vaccines in the United States: Recommendations of the National Vaccine Advisory Committee," *Journal of the American Medical Association* 290 (2003): 3122–3128.

7. Past Is Prologue

105 Rotting flesh: Tucker, *Scourge*, 2.
105 Smallpox disease and spread: Ibid., 2.
106 British historian: Quoted in Ibid., 3.
106 Smallpox deaths: Ibid., 3.
106 Monarchs and rulers died from smallpox: Ibid., 12.
106 Native Americans died from smallpox: Ibid., 12.
106 Edward Jenner and the smallpox vaccine: Ibid., 23.
107 Jenner's publication and acceptance of smallpox vaccine: Brunton, *Politics of Vaccination*, 13–14.
108 Epidemiological Society of London: Durbach, *Bodily Matters*, 22; Brunton, *Politics of Vaccination*, 40.
108 "Light of modern science": Durbach, *Bodily Matters*, 22.
108 Epidemiological Society's political role: Brunton, *Politics of Vaccination*, 40.
108 Epidemiological Society's reason for vaccination: Durbach, *Bodily Matters*, 23.
108 Bill of 1853: Brunton, *Politics of Vaccination*, 41.
108 Smallpox epidemic of 1852: Ibid.
108 "A damp squib": Ibid., 39.
108 Act of 1867: Durbach, *Bodily Matters*, 8–9.
109 "Ignorance and prejudice": Brunton, *Politics of Vaccination*, 43.
109 Richard Butler Gibbs and the Anti-Compulsory Vaccination League: Durbach, *Bodily Matters*, 38.
109 Growth of the ACVL: Ibid.
109 Formation of other anti-vaccination leagues: Ibid.
109 Richard Butler Gibbs quote regarding Britannia: Brunton, *Politics of Vaccination*, 92.
109 The *Vaccination Vampire*: Durbach, *Bodily Matters*, 138.
109 Vaccinators as ravens: Ibid.
109 Vaccination as sacrifice to the devil: Ibid., 81.
109 "Savage African tribe": Ibid.
110 Claim that vaccine contains animal products and transforms children into monsters: Durbach, *Bodily Matters*, 114.

110 Rallies at public auctions: Ibid., 53.

110 Rally in 1887: Ibid.

110 Rally in 1885: Ibid., 63.

110 Mothers hide children from vaccination officers: Ibid., 65.

111 John Gibbs regarding doctors: Ibid., 13.

111 Fisher and doctors killing babies: "Doctors Want Power to Kill Disabled Babies," posted by Barbara Loe Fisher, November 5, 2006, http://www.vaccineawakening.blogspot.com.

111 Leicester rally of 1885: Durbach, Bodily Matters, 51.

112 "Perfect carnival": Ibid., 50.

112 Anti-vaccine rally: Witnessed by the author, June 2006.

112 Bill of 1853 passed under darkness: Ibid., 118.

112 Fisher at Maryland courthouse: "Police with Dogs: Vaccinating Kids in Maryland," posted by Barbara Loe Fisher, November 19, 2007, http://www.vaccineawakening.blogspot.com.

113 Jane Orient on Nightline: "Vaccines and Their Risks," Nightline, October 14, 1999.

114 James Gillray's cartoon: Durbach, Bodily Matters, 124–125.

114 "Ox-faced boy or children": Brunton, Politics of Vaccination, 61.

114 "Horn-like excrescences": Durbach, Bodily Matters, 125.

114 Madhouses: Ibid.

114 "Low and browse": Ibid.

115 George Gibbs regarding smallpox vaccine: Ibid., 3.

115 Claim that smallpox vaccine causes children to change race: Ibid., 135.

115 Claim that smallpox vaccine causes diphtheria: Ibid., 183.

115 Claim that smallpox vaccine causes polio: H. Emerson, A Monograph on the Epidemic of Poliomyelitis (Infantile Paralysis) (New York: Arno Press, 1977).

116 "Right to be pure and unpolluted": Durbach, Bodily Matters, 71.

116 Avoiding smallpox: Ibid., 121.

116 McCarthy at "Green Our Vaccines" rally: Jenny McCarthy, "Green Our Vaccines" Rally, June 4, 2008, http://www.generationrescue.org/news-press/green-our-vaccines/jenny-mccarthy-transcript.

117 Frightened by germs: Durbach, Bodily Matters, 160.

118 "This infection scare is a sham": Ibid.

118 "The real enemy of the human race: dirt": Ibid., 164.

118 Early history of chiropractic: J. B. Campbell, J. W. Busse, and H. S. Injeyan, "Chiropractors and Vaccination: A Historical Perspective," Pediatrics, www.pediatrics.org/cgi/content/full/105/4/e43. All quotes from Daniel and B. J. Palmer were obtained from this source.

119 Nature paper: K. Wang, H. Zhang, M. Deqiong, et al., "Common Genetic Variants on 5p14.1 Associate with Autism Spectrum Disorders," Nature 459 (2009): 528–533.

120 Walter Hadwin regarding Elie Metchnikoff: Durbach, Bodily Matters, 168.

121   **"Pharaoh's daughter":** Durbach, *Bodily Matters*, 62.

121   **"Herodian decree":** Ibid.

121   **Vaccines and anti-Christ:** Ibid., 118.

122   ***Jenner or Christ?*:** Ibid.

122   **Mary Hume-Rothery:** Ibid.

122   **Debi Vinnedge:** Offit, *Vaccinated*, 90.

122   **Vinnedge and the Pontifical Academy for Life:** Ibid., 91.

122   **Mass-marketing of anti-vaccine message in Victorian England:** Durbach, *Bodily Matters*, 47-48.

123   **Ernest Hart regarding anti-vaccine rhetoric:** Ibid., 50.

123   **Rahul Parikh and modern-day messaging:** R. K. Parikh, "Fighting for the Reputation of Vaccines: Lessons from American Politics," *Pediatrics* 121 (2008): 621-622.

124   **Working-class resistance to vaccination:** Durbach, *Bodily Matters*, 92.

124   **Socioeconomic background of parents who refuse vaccines:** P. J. Smith, S. Y. Chu, and L. E. Barker, "Children Who Have Received No Vaccines: Who Are They and Where Do They Live?" *Pediatrics* 114 (2004): 187-195.

124   **Fisher and Kerensky:** National Vaccine Information Center, "President Bush Extends Filing Deadline on Compensation for Parents Under the National Childhood Vaccine Injury Fund," *Southwest Newswire*, November 8, 1990.

124   **Links to personal-injury lawyers:** www.nvic.org, April 2010.

125   **Conscientious-objection law:** Durbach, *Bodily Matters*, 171.

125   **Vaccination rates in England, late 1890s:** Ibid., 10.

127   **Vaccination rates in England versus Ireland and Scotland:** Brunton, *Politics of Vaccination*, 122-162.

8. Tragedy of the Commons

127   **"The most important Supreme Court case":** Gostin, *Public Health Law*, 116.

127   **Cited in other Supreme Court decisions:** Ibid., 125.

127   **Smallpox outbreak in Boston:** L. O. Gostin, "*Jacobson v. Massachusetts* at 100 Years: Police Power and Civil Liberties in Tension," *American Journal of Public Health* 95 (2005): 576-581.

127   **Cambridge ordinance:** Gostin, *Public Health Law*, 118.

128   **Vaccination vs. salvation:** "Editorial Points," *Boston Daily Globe*, November 19, 1901.

128   **Smallpox epidemic in Massachusetts:** W. E. Parmet, R. A. Goodman, and A. Farber, "Individual Rights vs. the Public Health—100 Years After *Jacobson v. Massachusetts*," *New England Journal of Medicine* 352 (2005): 652-654.

128   **Edwin Spencer:** Colgrove, *State of Immunity*, 38.

128   **Life of Henning Jacobson:** Ibid., 38-39; W. E. Parmet, R. A. Goodman, and A. Farber, "Individual Rights vs. the Public Health—100 Years After

Jacobson v. Massachusetts," *New England Journal of Medicine* 352 (2005): 652–654.

128 **Jacobson's case in local courts:** Colgrove, *State of Immunity*, 40.

128 **Henry Ballard and James Pickering:** W. E. Parmet, R. A. Goodman, and A. Farber, "Individual Rights vs. the Public Health—100 Years After *Jacobson v. Massachusetts,*" *New England Journal of Medicine* 352 (2005): 652–654.

129 **"Sacred Cow":** Ibid., 654.

129 **George Williams:** Colgrove, *State of Immunity*, 41.

129 **Williams's brief regarding civil liberty:** Gostin, *Public Health Law*, 121.

129 **Williams's brief regarding filth and disease:** W. E. Parmet, R. A. Goodman, and A. Farber, "Individual Rights vs. the Public Health—100 Years After *Jacobson v. Massachusetts,*" *New England Journal of Medicine* 352 (2005): 652–654.

130 **Harlan ruling:** *Jacobson v. Massachusetts*, 197 U.S. 11 (1905).

130 **Harlan regarding social compact:** Colgrove, *State of Immunity*, 42.

130 **Zucht ruling:** *Zucht v. King*, 260 U.S. 174 (1922).

130 **Mary Mallon:** The story of Mary Mallon, including all details and quotes, can be found in Leavitt, *Typhoid Mary*.

133 **Editorial on *Jacobson v. Massachusetts*:** *New York Times*, February 22, 1905.

133 **McCauley incident:** Colgrove, *State of Immunity*, 22.

134 ***New York Times* on McCauley incident:** "Quarantined Family Escapes," *New York Times*, March 23, 1894.

134 **Brooklyn Compulsory Anti-Vaccination League:** Colgrove, *State of Immunity*, 26.

134 **Massachusetts Compulsory Anti-Vaccination Association:** Ibid., 41.

134 **Anti-Vaccination League of America:** Ibid., 52.

134 **John Pitcairn regarding tyranny:** Ibid., 52.

134 **Anti-Vaccination League of America pamphlets:** Ibid., 54.

134 **Citizens' Medical Reference Bureau:** Ibid., 54–55.

134 **Lora Little and the American Medical Liberty League:** Allen, *Vaccine*, 99; Colgrove, *State of Immunity*, 56.

135 **Lora Little and "foul business":** Colgrove, *State of Immunity*, 60–61.

135 **Raggedy Ann doll:** Allen, *Vaccine*, 99.

136 **Decline of anti-vaccine groups:** Colgrove, *State of Immunity*, 74.

136 **CDC launches measles vaccine initiative:** Ibid., 149.

136 **Measles:** W. A. Orenstein and A. R. Hinman, "The Immunization System in the United States—The Role of School Immunization Laws," *Vaccine* 17 (1999): S19–S24.

136 **U.S. measles cases in early 1970s:** Colgrove, *State of Immunity*, 166–167.

136 **Joseph P. Kennedy Foundation:** Ibid., 175–176.

137 **Measles and encephalitis:** Centers for Disease Control and Prevention, "Update: Measles—United States, January–July 2008," *Morbidity and Mortality Weekly Report* 57 (2008): 893–896.

137 **Betty Bumpers and the Childhood Immunization Initiative:** Colgrove, *State of Immunity*, 199–200.

138 **Increased number of states requiring vaccines:** Ibid., 177.

138 **Measles outbreak in Alaska:** W. A. Orenstein and A. R. Hinman, "The Immunization System in the United States—The Role of School Immunization Laws," *Vaccine* 17 (1999): S19–S24.

138 **Measles outbreak in Los Angeles County:** Ibid.

138 **Texarkana:** P. J. Landrigan, "Epidemic Measles in a Divided City," *Journal of the American Medical Association* 221 (1972): 567–570.

138 **School entry requirements in fifty states:** Colgrove, *State of Immunity*, 177.

138 **Incidence of measles in United States, 1998:** W. A. Orenstein and A. R. Hinman, "The Immunization System in the United States—The Role of School Immunization Laws," *Vaccine* 17 (1999): S19–S24.

139 **Orenstein regarding Los Angeles mandates:** Author interview with Walter Orenstein, December 18, 2009.

139 **Prince case:** *Prince v. Massachusetts*, 321 U.S. 158 (1944). Italics added for emphasis.

140 **Wright case:** *Wright v. DeWitt High School*, 385 S.W. 2d 644 (Ark. 1965).

140 **McCartney case:** *McCartney v. Austin*, 293 N.Y.S. 2d 188 (1968).

140 **Avard case:** *Avard v. Manchester Board of School Committee et al.*, 376 F. Supp. 479 (1974).

140 **Brown case:** *Brown v. Stone*, 378 So. 2d 218 (1979).

140 **Davis case:** *Davis v. Maryland*, 294 Md. 370 (1982).

140 **New York State bill regarding mandatory polio vaccine:** S. H. Schanberg, "Assembly Votes Polio-Shots Bill: Vaccination Would Be Made Compulsory for Pupils," *New York Times*, June 21, 1966.

141 **Mary Baker Eddy regarding smallpox:** Eddy, *Science and Health*, 153.

141 **Diphtheria in Christian Scientist child:** Fraser, *Perfect Child*, 303.

141 **Measles outbreak at Principia College:** Centers for Disease Control and Prevention, "Outbreak of Measles Among Christian Science Students—Missouri and Illinois, 1994," *Morbidity and Mortality Weekly Report*, July 1, 1994.

141 **Orenstein regarding measles outbreak among Christian Scientists:** R. Goodrich, "Test Results Awaited in Causes of Deaths in Measles Outbreak," *St. Louis Post-Dispatch*, March 4, 1985; Fraser, *Perfect Child*, 302.

141 **Polio in Connecticut:** Centers for Disease Control and Prevention, "Follow-Up on Poliomyelitis," *Morbidity and Mortality Weekly Report* 21 (1972): 365–366; F. M. Foote, G. Kraus, M. D. Andrews, and J. C. Hart, "Polio Outbreak in a Private School," *Connecticut Medicine*, December 1973.

141 **Health commissioner responds to polio outbreak:** S. W. Ferguson, "Mandatory Immunization," *New England Journal of Medicine* 288 (1973): 800.

142 **Maier case:** *Maier v. Besser*, 73 Misc. 2d 241 (1972).

142 **Dalli case:** *Dalli v. Board of Education*, 358 Mass. 753 (1971).

142 **Sherr/Levy case:** *Sherr v. Northport–East Northport Union Free School District*, 672 F. Supp. 81 (E.D.N.Y 1987).

142 **Philosophical exemptions:** *Welsh v. United States*, 398 U.S. 333, 90 S.Ct. 1792 (1970).

143 **Salmon study:** D. A. Salmon, M. Haber, E. J. Gangarosa, et al., "Health Consequences of Religious and Philosophical Exemptions from Immunization Laws: Individual and Societal Risk of Measles," *Journal of the American Medical Association* 281 (1999): 47–53.

143 **Feikin study:** D. R. Feikin, D. C. Lezotte, R. F. Hamman, et al., "Individual and Community Risks of Measles and Pertussis Associated with Personal Exemptions to Immunization," *Journal of the American Medical Association* 284 (2000): 3145–3150.

143 **Omer study:** S. B. Omer, W.K.Y. Pan, N. A. Halsey, et al., "Nonmedical Exemptions to School Immunization Requirements: Secular Trends and Association of State Policies with Pertussis Incidence," *Journal of the American Medical Association* 296 (2006): 1757–1763.

143 **Glanz study:** J. M. Glanz, D. L. McClure, D. J. Magid, et al., "Parental Refusal of Pertussis Vaccination Is Associated with an Increased Risk of Pertussis Infection in Children," *Pediatrics* 123 (2009): 1446–1451.

144 **International impact of anti-vaccine activism:** E. J. Gangarosa, A. M. Galazka, C. R. Wolfe, et al., "Impact of Anti-Vaccine Movements on Pertussis Control: The Untold Story," *The Lancet* 351 (1998): 356–361.

144 **Garrett Hardin essay:** G. Hardin, "The Tragedy of the Commons," *Science* 162 (1968): 1243–1248. All Hardin quotes are from this essay.

146 **Stephanie Tatel:** S. Tatel, "A Pox on You," http://www.slate.com/toolbar .aspx?action=print&id=2232977. All Tatel quotes are from this essay.

147 **Hardin's second essay:** G. Hardin, "Extension of 'The Tragedy of the Commons,'" *Science* 280 (1998): 682–683.

9. The Mean Season

150 **Jenny McCarthy and Crystal children:** J. McCarthy, *Louder Than Words: A Mother's Journey in Healing Autism* (New York: Dutton, 2007), 178.

150 **Crystal children:** D. Virtue, *The Care and Feeding of Indigo Children* (Carlsbad: Hay House, 2001).

151 **McCarthy regarding MMR causing autism:** *Oprah*, September 18, 2007.

151 **McCarthy dramatizes son's illness:** K. Chetry, "McCarthy Claims Autism 'Cure,'" *American Morning*, CNN, October 1, 2008.

151 **McCarthy asks officials and companies to stop hurting kids:** Ibid.

152 **McCarthy on toxins in vaccines:** Green Our Vaccines Rally, Washington, D.C., June 4, 2008, http://www.generationrescue.org/news-press/green -our-vaccines/jenny-mccarthy-transcript.

152   **McCarthy on Botox:** S. Negovan, "The New McCarthyism," *Michigan Avenue Magazine,* http://www.michiganavemag.com/MA_MA09_084 _NEW.html.

152   **Jeffrey Kluger interviews McCarthy:** J. Kluger, "Jenny McCarthy Talks About Autism," *Time,* April 1, 2009.

153   **McCarthy video on treating autism:** http:///www.youtube.com/watch ?v=nDe_PAltC1A; http://www.youtube.com/watch?v=0mBqta02d68& feature=related. Other McCarthy quotes are from this video.

153   **Relative sizes of vaccine and nutritional supplement industries:** J. Groopman, "No Alternative," *Wall Street Journal,* August 7, 2006; "Infectious Diseases Vaccine Market Overview: Key Companies and Strategies," www.datamonitor.com, December 2008.

154   **McCarthy and Generation Rescue Web site:** "Jenny McCarthy and Jim Carrey Discuss Autism: Medical Experts Weigh In," *Larry King Live,* CNN, April 3, 2009.

154   **Chelation death:** K. Kane and V. Linn, "Boy Dies During Autism Treatment," *Pittsburgh Post-Gazette,* August 25, 2006.

154   **Generation Rescue mission statement:** Generation Rescue, Inc., 2007 Return of Organization Exempt from Income Tax, Form 990.

154   **Ad regarding mercury poisoning:** *New York Times,* June 8, 2005.

154   **Ad regarding increase in autism:** *USA Today,* April 6, 2006.

154   **Ad regarding Bailey Banks:** *USA Today,* February 25, 2009.

155   **Handley/McCarthy interchange with Dr. Travis Stork:** *The Doctors,* May 6, 2009.

156   **Handley and "knuckleheads":** J. B. Handley, "Dr. Steve Novella: Why Is This So Hard to Understand?" Age of Autism Web log, April 22, 2009.

157   **David Tayloe's letter to Lisa Williams:** Provided by Susan Martin, American Academy of Pediatrics.

158   **Nancy Minshew's quote in newspaper:** M. Roth, "Pitt Expert Goes Public to Counter Fallacy on Autism," *Pittsburgh Post-Gazette,* January 31, 2008.

158   **Handley counters Minshew:** Blog entry, Age of Autism, "While Minshew Shrieks, Autism Squeaks," February 8, 2008.

158   **Amanda Peet and *Cookie* magazine:** J. Tung, "Amanda Peet: The Actress Discusses the Vaccination Debate," *Cookie,* August 2008.

158   **Handley counters Peet:** Blog entry, Age of Autism, "From Dr. Paul Offit's Lips to Amanda Peet's Ears, We're All Parasites," July 15, 2008.

158   **Handley attacks ECBT:** Blog entry, "Every Child by Two: A Front Group for Wyeth," August 4, 2008.

159   **Handley counters Dawson:** Blog entry, Age of Autism, "Is Autism Speaks' Geri Dawson a Blithering Idiot?" September 10, 2008.

159   **Handley counters Snyderman:** Blog entry, Age of Autism, "Keep On Self-Incriminatin'," October 30, 2008.

159   **Harris article:** G. Harris and A. O'Connor, "On Autism's Causes: It's Parents vs. Research," *New York Times,* June 25, 2005.

159 **Handley counters Harris:** Blog entry, Age of Autism, "Some *New York Times* Reporters Are Just Ignorant," December 15, 2008.

159 ***Wired* article:** A. Wallace, "An Epidemic of Fear," *Wired*, November 2009.

159 **Handley and date rape:** "Readers Respond to An Epidemic of Fear: Part 1," *Wired*, http://www.wired.com/magazine/2009/10/readers-respond-to-an-epidemic-of-fear-part-1.

160 **Handley and "cry baby":** "Readers Respond to An Epidemic of Fear: Part 2," *Wired*, http://www.wired.com/magazine/2009/10/readers-respond-to-an-epidemic-of-fear-part-2.

160 **Wallace regarding bullying:** Ibid.

160 **Wallace and civil debate:** M. Block, "Journalist's Vaccine Article Draws Hate Mail," *National Public Radio*, October 28, 2009.

160 **Handley regarding pediatricians:** Blog entry, Age of Autism, "Vaccines Don't Cause Autism, Pediatricians Do," January 12, 2010.

160 **Handley and Gardasil:** "Jenny McCarthy and Jim Carrey Discuss Autism: Medical Experts Weigh In," *Larry King Live*, CNN, April 3, 2009.

160 **Patricia Danzon on vaccine economics:** Presentation on "Vaccine Economics," the University of Pennsylvania School of Medicine, Vaccines and Immune Therapeutics course for immunology and molecular biology graduate students, October 2008.

161 **Handley and the 1989 vaccine schedule:** *The Doctors*, May 6, 2009.

161 **Handley and research on vaccinated vs. unvaccinated children:** "Jenny McCarthy and Jim Carrey Discuss Autism: Medical Experts Weigh In," *Larry King Live*, CNN, April 3, 2009.

162 **Celebrities support polio vaccine:** N. G. Seavey, J. S. Smith, and P. Wagner, *A Paralyzing Fear: The Triumph over Polio in America* (New York: TV Books, 1998); Oshinsky, *Polio*.

162 **Celebrities who support vaccines:** "Actress and New Mom Keri Russell Joins PKIDS to Launch 'Silence the Sound of Pertussis' Campaign," *PR Newswire*, October 16, 2007; E. J. Mundell, "Jennifer Garner Puts Flu Shot in the Spotlight," *U.S. News and World Report*, www.usnews.com, April 19, 2010; "Football Legend Archie Griffin Raising Awareness of H1N1 Vaccinations," *Dayton Business Journal*, January 20, 2010.

163 **Celebrities who fear vaccines:** "Jessica Alba Concerned About Vaccinations," http://blogs.babycenter.com/celebrities/category/vaccinations; "Vaccines: Does Your Child Really Need Them?" Planet Chiropractic.com, April 18, 2000; "House Call with Dr. Sanjay Gupta," CNN, August 16, 2008; "Vaccines and Autism," *Larry King Live*, CNN, February 27, 2008; A. Quinn, "Aidan Quinn's Ode to Ireland of the Seventies," *Independent.ie*, September 14, 2008.

163 **Carrey regarding conspiracy:** "Jenny McCarthy and Jim Carrey Discuss Autism: Medical Experts Weigh In," *Larry King Live*, CNN, April 3, 2009. All quotes by Jim Carrey are from this source.

164   Handley and the chickenpox vaccine: Ibid.

164   Chickenpox disease and effectiveness of the chickenpox vaccine: A. A. Gershon, M. Takahashi, and J. F. Seward, "Varicella Vaccine," in *Vaccines*, 5th ed., eds. S. A. Plotkin, W. A. Orenstein, and P. A. Offit (London: Elsevier/Saunders, 2008); S. S. Shah, S. M. Wood, X. Luan, and A. J. Ratner, "Decline in Varicella-Related Ambulatory Visits and Hospitalizations in the United States Since Routine Immunization Against Varicella," *Pediatric Infectious Disease Journal* 29 (2010): 199–204.

165   Maher and Frist: *Real Time with Bill Maher*, HBO, October 2, 2009.

165   Maher in *Huffington Post*: B. Maher, "Vaccination: A Conversation Worth Having," *Huffington Post*, November 15, 2009.

166   *Religulous*: Lionsgate Films, 2008.

167   Maher and *Microbe Hunters*: B. Maher, "Vaccination: A Conversation Worth Having," *Huffington Post*, November 15, 2009.

167   Polio epidemics: Offit, *Cutter*.

167   Swine flu epidemic, 2009: Centers for Disease Control and Prevention, http://www.cdc.gov/h1n1flu/update.htm.

167   Swine flu and pregnancy: S. Jain, L. Kamimoto, A. M. Bramley, et al., "Hospitalized Patients with 2009 H1N1 Influenza in the United States, April–June 2009," *New England Journal of Medicine* 361 (2009): 1935–1944; J. K. Louie, M. Acosta, D. J. Jamieson, et al., "Severe 2009 H1N1 Influenza in Pregnant Women in California," *New England Journal of Medicine* 362 (2010): 27–35; Centers for Disease Control and Prevention, "2009 Pandemic Influenza A (H1N1) in Pregnant Women Requiring Intensive Care—New York City, 2009," *Morbidity and Mortality Weekly Report* 59 (2010): 321–326.

168   Edward R. Murrow and Arthur Godfrey: A. M. Brandt, *The Cigarette Century: The Rise, Fall, and Deadly Persistence of the Product That Defined America* (New York: Basic Books, 2007).

168   Bernadine Healy on TV: *CBS Evening News*, May 12, 2008. Other quotes from Healy are taken from this program.

169   Genetic defects in autism: K. Wang, H. Zhang, D. Ma, et al., "Common Genetic Variants on 5p14.1 Associate with Autism Spectrum Disorders," *Nature* 459 (2009): 528–533; R. Moessner, C. R. Marshall, J. S. Sutcliffe, et al., "Contribution of *SHANK-3* Mutations to Autism Spectrum Disorder," *American Journal of Human Genetics* 81 (2007): 1289–1297; K. Garber, "Autism's Cause May Reside in Abnormalities of the Synapse," *Science* 317 (2007): 190–191; L. A. Weiss, Y. Shen, J. M. Korn, et al., "Association Between Microdeletion and Microduplication at 16p11.2 and Autism," *New England Journal of Medicine* 358 (2008): 667–675.

169   Structural differences: "A Dose of Controversy," *Dateline NBC*, August 30, 2009.

169   Studies evaluating mercury toxicity in vaccines: Institute of Medicine, *Immunization Safety Review: Vaccines and Autism* (Washington, D.C.: National Academies Press, 2004).

10. Dr. Bob

171 **Sears's book:** R. W. Sears, *The Vaccine Book: Making the Right Decision for Your Child* (New York: Little, Brown, 2007). All quotes from Robert Sears were obtained from this source.

172 **Parents want alternative schedule:** Author interview with several pediatricians.

173 **Number of immunological components in vaccines:** P. A. Offit, J. Quarles, M. A. Gerber, et al., "Addressing Parents' Concerns: Do Multiple Vaccines Overwhelm or Weaken the Infant's Immune System?" *Pediatrics* 109 (2002): 124–129.

174 **Number of viral infections in first few years of life:** J. H. Dingle, G. F. Badger, W. S. Jordan, *Illness in the Home: A Study of 25,000 Illnesses in a Group of Cleveland Families* (Cleveland: The Press of Western Reserve University, 1964).

174 **Cohn and Langman paper:** M. Cohn and R. E. Langman, "The Protecton: The Unit of Humoral Immunity Selected by Evolution," *Immunological Reviews* 115 (1990): 9–147.

175 **Vaccine spacing and neurological outcomes:** M. J. Smith and C. R. Woods, "On-Time Vaccine Receipt in the First Year Does Not Adversely Affect Neuropsychological Outcomes," *Pediatrics* 125 (2010): 1–8.

175 **Hepatitis B vaccine studies on newborns:** S. M. Wheely, P. T. Jackson, E. H. Boxhall, et al., "Prevention of Perinatal Transmission of Hepatitis B Virus (HBV): A Comparison of Two Prophylactic Schedules," *Journal of Medical Virology* 35 (1991): 212–215; V. C. Wong, H. M. Ip, H. W. Reesink, et al., "Prevention of the HBsAg Carrier State in Newborns of Mothers Who Are Chronic Carriers of HBsAg and HBeAg by Administration of Hepatitis-B Vaccine and Hepatitis-B Immunoglobulin: Double-Blind Randomized Placebo-Controlled Study," *The Lancet* 28 (1984): 921–926; O. W. Prozesky, C. E. Stevens, W. Szmuness, et al., "Immune Response to Hepatitis B Vaccine in Newborns," *Journal of Infection* 7 (suppl. I) (1983): 53–55.

177 **Studies on aluminum elimination and toxicity:** N. W. Baylor, W. Egan, and P. Richman, "Aluminum Salts in Vaccines—US Perspective," *Vaccine* 20 (2002): S18–S23; N. J. Bishop, R. Morley, J. P. Day, and A. Lucas, "Aluminum Neurotoxicity in Preterm Infants Receiving Intravenous-Feeding Solutions," *New England Journal of Medicine* 336 (1997): 1557–1561; Committee on Nutrition, "Aluminum Toxicity in Infants and Children," *Pediatrics* 97 (1996): 413–416; P. O. Ganrot, "Metabolism and Possible Health Effects of Aluminum," *Environmental Health Perspectives* 65 (1986): 363–441; L. S. Keith, D. E. Jones, and C. Chou, "Aluminum Toxicokinetics Regarding Infant Diet and Vaccinations," *Vaccine* 20 (2002): S13–S17; J. A. Pennington, "Aluminum Content of Food and Diets," *Food Additives and Contaminants* 5 (1987): 164–232; K. Simmer, A. Fudge, J. Teubner, and S. L. James, "Aluminum

Concentrations in Infant Formula," *Journal of Paediatrics and Child Health* 26 (1990): 9–11.

179 **Formaldehyde:** "Epidemiology of Chronic Occupational Exposure to Formaldehyde: Report of the Ad Hoc Panel on Health Aspects of Formaldehyde," *Toxicology and Industrial Health* 4 (1988): 77–90; A. T. Natarajan, F. Darroudi, C.J.M. Bussman, et al., "Evaluation of the Mutagenicity of Formaldehyde in Mammalian Cytogenetic Assays *In Vivo* and *In Vitro*," *Mutation Research* 122 (1983): 355–360; H. P. Til, R. A. Woutersen, V. J. Feron, et al., "Two-Year Drinking-Water Study of Formaldehyde in Rats," *Food Chemical Toxicology* 27 (1989): 77–87; F. M. Huennekens and M. J. Osborne, "Folic Acid Coenzymes and One-Carbon Metabolism," *Advances in Enzymology* 21 (1959): 369–446; H. Heck, M. Casanova-Schmitz, P. B. Dodd, et al., "Formaldehyde (CH_2O) Concentrations in the Blood of Humans and Fischer-344 Rats Exposed to CH_2O Under Controlled Conditions," *American Industrial Hygiene Association Journal* 46 (1985): 1–3.

180 **VAERS reports by personal-injury lawyers:** M. J. Goodman and J. Nordin, "Vaccine Adverse Events Reporting System Reporting Source: A Possible Source of Bias in Longitudinal Studies," *Pediatrics* 117 (2006): 387–390.

181 **MMR vaccine causes a low platelet count:** R. A. Oski and J. L. Naiman, "Effect of Live Measles Vaccine on the Platelet Count," *New England Journal of Medicine* 275 (1966): 352–356.

181 **Thimerosal in vaccines doesn't cause autism:** K. M. Madsen, M. B. Lauritsen, C. B. Pedersen, et al., "Thimerosal and the Occurrence of Autism: Negative Ecological Evidence from Danish Population-Based Data," *Pediatrics* 112 (2003): 604–606; A. Hviid, M. Stellfeld, J. Wohlfahrt, and M. Melbye, "Association Between Thimerosal-Containing Vaccine and Autism," *Journal of the American Medical Association* 290 (2003): 1763–1766.

181 **HPV vaccine testing:** J. T. Schiller, I. H. Frazer, and D. R. Lowy, "Human Papillomavirus Vaccines," in *Vaccines*, 5th ed., eds. S. A. Plotkin, W. A. Orenstein, and P. A. Offit (London: Elsevier/Saunders, 2008).

181 **Pneumococcal vaccine testing:** S. Black, H. Shinefield, B. Fireman, et al., "Efficacy, Safety and Immunogenicity of Heptavalent Pneumococcal Conjugate Vaccine in Children: Northern California Kaiser Permanente Vaccine Study Center Group," *Pediatric Infectious Disease Journal* 19 (2000): 187–195.

181 **Rotavirus vaccine testing:** T. Vesikari, D. O. Matson, P. Dennehy, et al., "Safety and Efficacy of a Pentavalent Human-Bovine (WC3) Reassortant Rotavirus Vaccine," *New England Journal of Medicine* 354 (2006): 22–33; G. M. Ruiz-Palacios, I. Perez-Schael, F. R. Velázquez, et al., "Safety and Efficacy of an Attenuated Vaccine Against Severe Rotavirus Gastroenteritis," *New England Journal of Medicine* 354 (2006): 11–21.

181 **Pneumococcal infections in children:** Centers for Disease Control and Prevention, "Preventing Pneumococcal Disease Among Infants and Young Children: Recommendations of the Advisory Committee on Immunization Practices (ACIP)," *Morbidity and Mortality Weekly Report* 49 (2000): 1–35.

182 **Shannon Peterson:** L. Szabo, "Missed Vaccines Weaken 'Herd Immunity' in Children," *USA Today*, January 6, 2010.

182 **Anti-vaccine activists and adders, bats:** Durbach, *Bodily Matters*, 114.

182 **McCarthy and vaccine ingredients:** http:///www.youtube.com/watch ?v= nDe_PAltC1A; http://www.youtube.com/watch?v=0mBqta02d68& feature=related.

182 **Prion biology:** P. A. Offit, R. L. Davis, and D. Gust, "Vaccine Safety," in *Vaccines*, 5th ed., eds. S. A. Plotkin, W. A. Orenstein, and P. A. Offit (London: Elsevier/Saunders, 2008).

183 **Hepatitis B infections in children:** G. L. Armstrong, E. F. Mast, M. Wojczynski, and H. S. Margolis, "Childhood Hepatitis B Virus Infections in the United States Before Hepatitis B Immunization," *Pediatrics* 108 (2001): 1123–1128.

186 **Diphtheria in the Russian Federation:** Centers for Disease Control and Prevention, "Diphtheria Outbreak—Russian Federation, 1990–1993," *Morbidity and Mortality Weekly Report* 42 (1993): 840–847.

186 **Hib in the United States:** Centers for Disease Control and Prevention, "Invasive *Haemophilus Influenzae* Type B Disease in Five Young Children—Minnesota, 2008," *Morbidity and Mortality Weekly Report* 58 (2008): 1–3; D. Sapatkin, "A Fatal Link in Vaccine Shortage," *Philadelphia Inquirer*, April 1, 2009.

188 **Mehmet Oz on TV:** *The Joy Behar Show*, January 12, 2010.

188 **Oz and Roizen book:** M. Oz and M. Roizen, *YOU: Having a Baby: The Owner's Manual to a Happy and Healthy Pregnancy* (New York: Simon & Schuster, 2009). All subsequent references to Oz's and Roizen's view of vaccines is taken from this source.

190 **FDA warning about rotavirus vaccine:** Food and Drug Administration, "FDA Public Health Notification: Information on RotaTeq and Intussusception," February 13, 2007.

190 **CDC study on rotavirus vaccine and intussusception:** P. Haber, M. Patel, H. S. Izurieta, et al., "Post-Licensure Monitoring of Intussusception After RotaTeq Vaccination in the United States, February 1, 2006–September 25, 2007," *Pediatrics* 121 (2008): 1206–1212.

192 **Chen graph:** R. T. Chen, et al., "The Natural History of an Immunization Program," *Vaccine* 12 (1994): 542–550.

11. Trust

193 **Christian Scientists and public responsibility:** Fraser, *Perfect Child*, 177–178.

193 Whitney case: Ibid., 271–272.
193 Sutherland case: Ibid., 276–277.
194 Cornelius case: Ibid., 277–279.
194 Sheridan case: Ibid., 279–284.
194 Swan case: Ibid., 285–298.
194 Walker case: Ibid., 298–300.
195 Twitchell case: Ibid., 303–305.
195 King case: Ibid., 305–309.
195 McKown case: Ibid., 310–317.
195 States repealing religious exemptions: Ibid., 337.
196 Hinman regarding exemptions: Author interview with Alan Hinman, December 7, 2009.
196 Orenstein regarding exemptions: Author interview with Walter Orenstein, December 18, 2009.
196 Influenza virus in hospitals: G. A. Poland, P. Tosh, and R. M. Jacobson, "Requiring Influenza Vaccination for Healthcare Workers: Seven Truths We Must Accept," *Vaccine* 23 (2005): 2251–2255.
197 Hospitals mandate influenza vaccine: T. Redlup, "Hospital Workers Fired After Refusing Flu Vaccine," *Vaccine News Daily*, January 14, 2010, http://vaccinenewsdaily.com/news/211617-hospital-workers-fired-after-refusing-flu-vaccine.
197 Influenza vaccine policy at the Children's Hospital of Philadelphia: E. Smith, "At CHOP, No Flu, No Job. No Sense?" *Philadelphia Daily News*, December 5, 2009.
198 Brad Dyer and manifesto: C. K. Johnson, "1 in 4 Parents Believe Vaccines Cause Autism," *Associated Press*, March 1, 2010.
198 Dyer's vaccine policy: B. J. Dyer, R. C. Duncheskie, A. Lehovich, et al., "All Star Pediatrics' Vaccine Policy Statement," *AAP News* 29 (2008): 26.
199 Dyer on parent response: C. K. Johnson, "1 in 4 Parents Believe Vaccines Cause Autism," *Associated Press*, March 1, 2010.
199 Salamone on CDC officials: Author interview with John Salamone, December 4, 2009.
200 Orenstein on being human: Author interview with Walter Orenstein, December 18, 2009.
200 Attkisson regarding Every Child by Two: "Follow the Money," Sharyl Attkisson and Katie Couric, *CBS Evening News*, July 25, 2008.
201 Attkisson regarding AAP: Ibid.
203 Orenstein regarding conspiracy theories: Author interview with Walter Orenstein, December 18, 2009.
203 David Aaronovitch regarding conspiracy theories: D. Aaronovitch, "A Conspiracy-Theory Theory," *Wall Street Journal*, December 19–20, 2009.
204 Rebecca Estepp on TV: *World News Tonight*, ABC, March 12, 2010.
204 Estepp in *New York Times*: D. G. McNeil, Jr., "3 Rulings Find No Link to Vaccines and Autism," *New York Times*, March 12, 2010.
205 Penny Heaton speech: Author witnessed this presentation in 2005.

Epilogue

207 **NPR program:** "Ruining It for the Rest of Us," *This American Life*, National Public Radio, December 19, 2008, www.thisamericanlife.org/ Radio_Episode.aspx?sched=1275.

210 **Yarkin:** Author interview with Celina Yarkin, February 10, 2010.

214 **Flint:** E. Carlyle, "Rare Hib Disease Increases in Minnesota: Is the Anti-Vaccine Movement to Blame?" *City Pages: The News and Arts Weekly of the Twin Cities*, June 3, 2009; L. Szabo, "Missed Vaccines Weaken 'Herd Immunity' in Children," *USA Today*, January 6, 2010.

SELECTED BIBLIOGRAPHY

Allen, Arthur. *Vaccine: The Controversial Story of Medicine's Greatest Lifesaver.* New York: W. W. Norton, 2007.

Benson, T. W., and C. Anderson. *Reality Fictions: The Films of Frederick Wiseman*, 2nd ed. Carbondale and Edwardsville: Southern Illinois University Press, 2002.

Boyce, Tammy. *Health, Risk and News: The MMR Vaccine and the Media.* New York: Peter Lang, 2007.

Brunton, Deborah. *The Politics of Vaccination: Practice and Policy in England, Wales, Ireland, and Scotland, 1800–1874.* Rochester, NY: University of Rochester Press, 2008.

Cather, Willa, and Georgine Milmine, *The Life of Mary Baker Eddy & the History of Christian Science.* Lincoln: University of Nebraska Press, 1993.

Colgrove, James. *State of Immunity: The Politics of Vaccination in Twentieth-Century America.* Berkeley: University of California Press, 2006.

Durbach, Nadja. *Bodily Matters: The Anti-Vaccination Movement in England, 1853–1907.* Durham, N.C.: Duke University Press, 2005.

Eddy, Mary Baker. *Science and Health with Key to the Scriptures.* Boston: First Church of Christ, Scientist, 1875.

Fitzpatrick, Michael. *MMR and Autism: What Parents Need to Know.* London: Routledge, 2004.

———. *Defeating Autism: A Damaging Delusion.* London: Routledge, 2009.

Fraser, Caroline. *God's Perfect Child: Living and Dying in the Christian Science Church.* New York: Henry Holt and Company, 1999.

Goldacre, Ben. *Bad Science.* London: Fourth Estate, 2008.

Gostin, Lawrence. *Public Health Law: Power, Duty, Restraint.* Berkeley: University of California Press, 2008.

Kabat, Geoffrey. *Hyping Health Risks: Environmental Hazards in Daily Life and the Science of Epidemiology.* New York: Columbia University Press, 2008.

Leavitt, Judith. *Typhoid Mary: Captive to the Public's Health.* Boston: Beacon Press, 1996.

Offit, Paul. *The Cutter Incident: How America's First Polio Vaccine Led to the Growing Vaccine Crisis.* New Haven, Conn.: Yale University Press, 2005.

———. *Vaccinated: One Man's Quest to Defeat the World's Deadliest Diseases*. New York: Smithsonian Books, 2007.

———. *Autism's False Prophets: Bad Science, Risky Medicine, and the Search for a Cure*. New York: Columbia University Press, 2008.

Oshinsky, David. *Polio: An American Story*. Oxford and New York: Oxford University Press, 2005.

Schoepflin, Rennie B. *Christian Science on Trial: Religious Healing in America*. Baltimore: Johns Hopkins University Press, 2003.

Sommerville, C. John. *How the News Makes Us Dumb: The Death of Wisdom in an Information Society*. Downers Grove, Ill.: InterVarsity Press, 1999.

Tucker, Jonathan. *Scourge: The Once and Future Threat of Smallpox*. New York: Atlantic Monthly Press, 2001.

ACKNOWLEDGMENTS

I wish to thank T. J. Kelleher for his wisdom, humor, scientific knowledge, and deft editorial hand; Andrew Zack for guiding me through the narrow streets of the publishing business; Bojana Ristich and Christine Arden for their lessons on logic, style, and form; Erica Johnson for her research assistance; Patrick Fitzgerald, Susan Martin, Don Mitchell, John O'Brien, Carl Offit, Bonnie Offit, Emily Offit, and Will Offit for their careful reading of the manuscript; and Jennifer Bardwell, Jeff Bergelson, Samuel Berkovic, Matt Carey, James Cherry, Kristen Feemster, Mark Feinberg, David Gorski, Lawrence Gostin, Penny Heaton, Alan Hinman, Phil Johnson, Matthew Kronman, Gary Marshall, Charlotte Moser, Sheila Nolan, Glen Nowak, Walter Orenstein, Georges Peter, Larry Pickering, Amy Pisani, Stanley Plotkin, Lisa Randall, Ken Reibel, Lance Rodewald, Lucy Rorke-Adams, John Salamone, David Salisbury, William Schaffner, Jason Schwartz, Alison Singer, Michael Smith, Mike Stanton, Kirsten Thistle, Dan Thomasch, Tom Vernon, Deborah Wexler, Margaret Williams, and Karie Youngdahl for their recollections of the vaccine controversy as well as their expertise.

I would also like to acknowledge the bravery of Ira Flatow, John Hamilton, Gardiner Harris, Claudia Kalb, Ron Lin, Anita Manning, Chris Mooney, Anahad O'Connor, Jon Palfreman, Rahul Parikh, Amanda Peet, Ami Schmitz, Nancy Snyderman, Michael Specter, Travis Stork, John Stossel, Liz Szabo, Trine Tsouderos, and Amy Wallace for their willingness to stand up for the science of vaccine safety independent of the cost.

INDEX

AAP. *See* American Academy of Pediatrics (AAP)

Aaronovitch, David, 203–204

ACIP. *See* Advisory Committee on Immunization Practices (ACIP)

ACVL. *See* Anti-Compulsory Vaccination League (ACVL)

Adams, Lucy Rorke, 88–91

Advisory Commission on Childhood Vaccines, 82

Advisory Committee on Immunization Practices (ACIP), 80, 82

AIDS (Acquired Immune Deficiency Syndrome), 50–51

Alaska, measles outbreak, 138

Alba, Jessica, 163

Alexander, Hattie, 128

Allergic reactions, to vaccines, 59–60

Allopathic medicine, 117

Aloudat, Tammam, xvii

Alternative medicine, lure of, 119

Althen, Margaret, 86–87

Aluminum, in vaccines, 176–178

American Academy of Family Physicians, 20

American Academy of Pediatrics (AAP), 10, 20, 96, 157, 160, 201

American Experience (television program), 184

American Medical Association, 20, 22

American Medical Liberty League, 134, 136

Angstadt, Lauren, 91

Anthrax, 117

Antibodies, 174

Anti-Compulsory Vaccination Act, 143

Anti-Compulsory Vaccination League (ACVL), 109, 115

Anti-Vaccination League, 128

Anti-Vaccination League of America, 134, 136

Anti-vaccine movement/activists
  birth of modern American, 2–12
  CDC and, 199–200
  celebrities and, 156–157, 162–168
  choosing not to vaccinate, 144–145
  conspiracy theories and, 203–204
  differences between 19th-century and modern movements, 123–125
  DPT: Vaccine Roulette and, 2–7, 12
  in England, 13–23
  Healy and, 168–170
  lawyers and, 124, 154
  marketing strategy, 124–125
  mass marketing and, 122–123
  pharmaceutical companies and, 202–203
  rejection of germ theory, 117–119
  religious beliefs and, 121–122
  similarities between 19th-century and modern movements, 111–123
  smallpox vaccine and, 105, 109–111, 134–136
  themes, 111–123
  vaccine advocates and, 200–202

Antivaccine philosophy, Steiner and, xiii

Anti-vaccine rallies/protests, 110, 111–112, 116–117, 172, 179

Arizona
  measles outbreak, xvi
  pertussis outbreak, xiii

Arkansas, measles outbreak, xvi

Aschoff, Ludwig, 89

Asher, Evan, 150–151
Asher, John, 150, 151
Association of Parents of Vaccine-
    Damaged Children, 16, 36, 38
Attkisson, Sharyl, 168–170, 200–202
Autism
    cures advanced for, 101–102, 119,
        153
    McCarthy and, 150–154
    mercury in vaccines (thimerosal) and,
        96–98, 99, 100, 102, 169
    MMR vaccine and, 92–94, 97–98,
        99, 150–151
    Omnibus Autism Proceeding,
        97–104
    vaccines and, 83, 91–92
Autism Action Coalition, 96
Autism Speaks, 159
Avard, Ronald, 140
Azidothymidine, 51

Bacterial meningitis, 69
Baker, Josephine, 132
Ballard, Henry, 128–129
Banks, Bailey, 154
Baraff, Larry, 27–28, 45, 49
Bazell, Robert, xi
BCG (Bacillus of Calmette and Guérin)
    vaccine, 55–56
Behar, Joy, 188
Belkin, Lyla, 66, 67
Belkin, Michael, 67–68
Berg, Anna, 44
Berkovic, Samuel, 40–42, 44, 76–77
Bernier, Roger, 83
"Bill to Further Extend and Make
    Compulsory the Practice of
    Vaccination," 108
Blaylock, Russell, 166, 167
Bonthrone, Iris, 33
Bonthrone, John, 33
Bonthrone, Richard, 32–33
Bordetella pertussis, xiii, 25
Boston Daily Globe (newspaper), 128
Bradstreet, Jeff, 101–102
Brain damage, pertussis vaccine and,
    2–11, 13–16, 28–40
Brain Research Institute, 40
Brandt, Edward, 21–22
Bridgewater State Hospital for the
    Criminally Insane, 1

British Medical Association, 114
British Medical Journal (journal), 15–16,
    123
British Pediatric Association, 30
Brooklyn Compulsory Anti-Vaccination
    League, 134
Brown, Charles, 140
Bumpers, Betty, 136, 137, 158, 201
Bumpers, Dale, 137
Burns, James B., 53
Burton, Susan, 208, 209
Byers, Randolph, 14, 34
Byers, Vera, 100
Byline: Lea Thompson (television
    program), 11

California, measles outbreaks, xv–xvi,
    138, 207
Calmette, Albert, 55–56
Campbell, Meagan, 209
Canadian National Advisory Committee
    on Immunization, 30
Canby, Vincent, 1–2
Capizzano, Rose, 87
The Care and Feeding of Indigo Children
    (Virtue), 150
Carlson, Sybil, 208
Carrey, Jim, as anti-vaccine activist, 116,
    163–164, 168, 172, 179, 202
Carter, Jimmy, 43, 184
Carter, Rosalynn, 43, 137, 159, 201
CBS Evening News (television program),
    168, 170, 200–202
CDC. See Centers for Disease Control and
    Prevention (CDC)
Cedillo, Michelle, 97, 102
Cedillo, Theresa and Michael, 102
Celebrity, use by anti-vaccine movement,
    156–157, 162–168
Center of Excellence in Autism, 158
Centers for Disease Control and
    Prevention (CDC)
    criticism of Vaccine Roulette, 10
    on effects of pertussis, 46–47
    on mercury in vaccines, 96
    on pertussis outbreaks, xiii–xiv
    rotavirus vaccine and, 72–73
    school immunization requirements and,
        136
    study of pertussis vaccine-brain damage
        link, 31

as target of anti-vaccine groups,
    199–200
vaccine shortages and, 19–20
Chadwick, Nicholas, 94
Chambers, Hillary, 208
Chambers, Finley, 208
Chase, Sylvia, 65–66, 68
Chemicals, in vaccines, 178–179
Chen, Robert, 191–192
Cherry, James, 16–17, 31
Chickenpox, 164
Chickenpox vaccine, 75–76, 164
CHILD. See Children's Health Care Is a
    Legal Duty (CHILD)
Childhood Immunization Initiative, 137
Child Neurology Association, 31
Children of God for Life, 122
Children's Health Care Is a Legal Duty
    (CHILD), 194
Children's Hospital of Philadelphia, 88,
    197
Chiropractors, vaccines and, 118–119
Christian Scientists, religious exemption
    to vaccination requirements and,
    141–142, 193–195
Christopher, Warren, 195
Church of Human Life Science, 140
Citizens' Medical Reference Bureau, 134
Classen, Bart, 61–62, 69–70
Classen Immunotherapies, 61
Clostridium tetani, 26
Coalition for Vaccine Safety, 149
Cohn, Mel, 174
Colantoni, Anthony, 53
Committee on the Safety of Medicines, 50
Compulsory vaccination, 139
    anti-vaccine movement and, 134–136
    Jacobson v. Massachusetts, 127–130
    for smallpox, 108–109, 127–130
    social good vs. individual freedom and,
        130
    Typhoid Mary and, 130–133
    See also Mandatory vaccination
Concomitant-use studies, 187
Confessions of a Medical Heretic
    (Mendelsohn), 48
Connaught Laboratories, 19
Conscientious objector status, 125, 143
Conspiracy theory on vaccines, 51–52,
    203–204
Converse, Ben, 66

Conway, Homer and Chin-Caplan, 102
Cookie (magazine), 158
Cornelius, David, 193
Cornelius, Edward and Anne, 193
Corynebacterium diphtheriae, 25
"Courage in Science Award," 94
"The Cow-Pock or the Wonderful Effects
    of the New Inoculation" (Gillray),
    114
Cowpox, 106–107
Crawford, Cindy, 163
The Crime Against the School Child
    (Higgins), 134
Crimes of the Cowpox Ring (Little), 135,
    180, 183
Culturelle, 153
Cutter Laboratories, 55

Dalli, Belinda, 142
Dalli, Beulah, 142
Danzon, Patricia, 160
Dateline NBC (television program), 11
Daum, Robert, 71
Davis, Irving, 140
Dawson, Geri, 159
Deafness, DTP vaccine and, 62–63
Deer, Brian, 93
Delaware, pertussis outbreak, xiii
Department of Community Medicine,
    29
Department of Defense, 20
Department of Health and Human
    Services, 20, 21
Diabetes, vaccines and, 61–62, 64, 70
Dick, George, 15
Diphtheria, xviii, 25–26, 185–186, 193
Diphtheria vaccine, 26, 191
Dissatisfied Parents Together (DPT), 8,
    10, 23, 26. See also National
    Vaccine Information Center
District of Columbia, measles outbreak,
    xvi
Doctors
    attitude of anti-vaccine movement
        towards, 48, 81–82, 111
    Handley and, 160
    refusing care for unvaccinated children,
        197–199
The Doctors (television program),
    155–157, 161, 171
Downs, Hugh, 67

DPT. *See* Dissatisfied Parents Together (DPT)

DPT: *Vaccine Roulette* (television documentary), 2–7, 18, 25, 28, 43
criticism of, 10, 45–54
experts used in, 47–53

The Dr. Oz Show (television program), 188

Dravet, Charlotte, 41

Dravet's syndrome, 41–42

DTP vaccine, deafness and, 62–63. *See also* Diphtheria vaccine; Pertussis vaccine; Tetanus vaccine

Dunbar, Bonnie, 65–66

Dyer, Brad, 198

East Bay Waldorf School whooping cough outbreak, xiii

ECBT. *See* Every Child by Two (ECBT)

Eddy, Mary Baker, 141

Eggs, allergic reactions to vaccines and, 59

Ehrich, William, 89

Einstein, Albert, 89

Eldering, Grace, 25, 26

Eli Lilly, 55

Endotoxin, in pertussis vaccine, 29

England
antivaccine movement in, 13–23
conscientious-objection law and vaccination rates, 125
DTP vaccine injury cases, 32–40

"An Epidemic of Fear" (Wallace), 159

Epidemiological Society of London, 108

Epidemiology Intelligence Service, 204

Epiglottitis, 60–61

Epilepsy
causes of, 40–44
pertussis vaccine and risk of, 28–32

Epilepsy Research Center, 40

Estepp, Rebecca, 204

Every Child by Two (ECBT), 137, 158–159, 200–201

Expert witnesses, autism cases and, 99–102

"Extension of 'The Tragedy of the Commons'" (Hardin), 147

The Facts Against Compulsory Vaccination (pamphlet), 134

Federal Circuit Court of Appeals, 86, 87

Feikin, Daniel, 143

Fever, seizures and, 40

"Fighting for the Reputation of Vaccines" (Parikh), 123

Fisher, Barbara Loe, 57, 202
on changes in VICP's vaccine-injury-compensation table, 86
on chickenpox vaccine, 75
child's reaction to vaccination and founding of Dissatisfied Parents Together, 7–8, 26, 28
distrust of doctors, 81–82
Hawkins hearings and, 9
hepatitis B vaccine and, 65, 67, 68
on herd immunity, 76
on Hib vaccine, 61
on HPV vaccine, 73–75
Jenny McCarthy's activism compared to, 151–154, 162
Maher and, 166, 167
as media source of information, 149, 168
on natural infection, 76
paranoia and, 112–113
parental guilt over vaccination problems and, 77
photo, 60
on pneumococcal vaccine, 69, 70, 71
A Shot in the Dark, 57, 61, 81, 91, 180, 183
on vaccine advisory council, 70–71
on vaccine-autism link, 91–92
as vaccine safety activist, 58–59, 80–83
on validity of scientific studies, 82–83
Wakefield and, 94–96

Fisher, Christian, 7–8

Fisher, Margaret, 160

Flint, Brendalee, 212–214

Flint, Julieanna, 212–214

Flutie, Doug, 163

Foege, William, 47, 211

Food and Drug Administration, 27, 43, 61, 71, 73

Formaldehyde, in vaccines, 178–179

Fourteenth Amendment, compulsory vaccination and, 129–130

Fox, Rosemary, 16, 36, 38

Francis, Thomas, 59

Frankenstein (Shelley), 120

Frist, Bill, 165

Fundamentals of Anthroposophical Medicine (Steiner), xiii

Galvani, Luigi, 120
Gardasil, 160–161. *See also* Human papillomavirus (HPV) vaccine
Garner, Jennifer, 162
Garrow, Irene, 14
Gates Foundation's Global Health Program, xvii
Gaugert, Polly, 4
Geier, Mark, 29
Gelatin, allergic reactions to vaccines and, 59
General Medical Council, 94–95, 96
Generation Rescue, 96, 154–155
Genetic cause of encephalopathy, 42–44
Georgia, measles outbreak, xvi
Germ theory, rejection of, 117–119
Gibbs, George, 109, 115
Gibbs, John, 109
Gibbs, Richard Butler, 109
Gillray, James, 114–116
Glanz, Jason, 143
Glaxo, 33
Glaxo-SmithKline, 183
Godfrey, Arthur, 168
Golden, Gerald, 31
Golkiewicz, Gary, 98
Gordon, Jay, 166, 167
Graham, Michelle, 19
Grant, Jim, 4
Grant, Marge, 4, 43–44, 52
Grant, Scott, 4, 46
"Green Our Vaccines" rally, 116–117, 172, 179
Griffin, Archie, 162
Griffin, Marie, 30
Group Health Cooperative, 30
Gruelle, Johnny, 135–136
Gruelle, Marcella, 135
Guérin, Camille, 55

Hadwin, Walter, 120
Haemophilus influenzae type b (Hib) infection, 60–61, 186
Haemophilus influenzae type b (Hib) meningitis, xi–xii, 63, 186, 213–214
Haemophilus influenzae type b (Hib) vaccine, 60–64, 87

Halsey, Neal, 69
Handley, J. B., 154–162, 168, 202
attacks on Gardasil vaccine, 160–161
confrontations, 155–160
on *The Doctors*, 155–158, 161
on *Larry King Live*, 160, 161, 164
photo, 156
Hardin, Garrett, 144–145, 147
Harlan, John Marshall, 130
Harris, Burton, 195
Harris, Gardiner, 159
Hart, Ernest, 123
Harvard School of Public Health, 30
Hasidic Jews, mumps outbreak among, xvi–xvii
Hawaii
measles outbreak, xvi
repeal of religious exemption health laws, 195
Hawkins, Paula, 8–10, 18, 40, 43
Hazelhurst, Rolf and Angela, 102
Hazelhurst, Yates, 97, 99, 102
Healthcare workers, mandatory influenza vaccination for, 196–197
Healy, Bernadine, as anti-vaccine activist, 168–170
Heaton, Penny, 204–206
Hellström, Bo, 29
Hepatitis B infection, 54, 183–184
Hepatitis B vaccine, 64–68
infant immune response to, 175
multiple sclerosis and, 65–66, 68, 88
rheumatoid arthritis and, 87
SIDS and, 67–68
vaccine court rulings, 87, 88
Hepatitis B Vaccine Project, 67
Herd (population) immunity, xvii, 76
breakdown in, 145–147, 184–185, 207–214
Heteropathic medicine, 117
Hib. *See under Haemophilus influenzae* type b (Hib)
Higgins, Charles, 134, 136, 180
Hinman, Alan, 6, 136, 196
H1N1 virus, 76, 165, 167
Hodgson, Abraham Victor Obeng, 205
Horrors of Vaccination Exposed and Illustrated (Higgins), 134, 180
Hospital for Sick Children, 14, 33
House Subcommittee on Health and the Environment, 19

How to Raise a Healthy Child in Spite of Your Doctor (Mendelsohn), 48
HPV. *See under* Human papillomavirus (HPV)
Huffington Post, 165, 167
Human immunodeficiency virus (HIV), 51
Human papillomavirus (HPV) vaccine, 73–75, 160
  fear of technology used to make it, 120–121
  testing, 181
Human serum albumin, in MMR vaccine, 183
Hume-Rothery, Mary, 122
Hypotonic hyporesponsive episodes, pertussis vaccine and, 45
Hypotonic Hyporesponsive Syndrome, 28

"Ileal-Lymphoid-Nodular Hyperplasia, Non-Specific Colitis, and Pervasive Developmental Disorder in Children" (Wakefield), 92
Illinois
  measles outbreak, xvi
  pertussis outbreak, xiii
Immunization programs, history of, 191–192
Immunization rates
  British pertussis vaccine, 16, 17
  effect of conscientious objection law on, 125
  herd immunity and, xvii
Immunological challenge, of vaccines, 26, 173–174
Imus, Don and Deirdre, 152
Indiana, measles outbreak, xiv
Infantile spasms, pertussis vaccine and, 46
Infants, immune response of, 173–176
Influenza vaccine
  egg allergy and, 59
  H1N1, 165, 167
  mandating by hospital administrators, 196–197
  Oz on, 189
Informed Parents Against Vaccine-Associated Polio (IPAV), 79
Inquiry into the Causes and Effects of the Variolae Vaccinae (Jenner), 107

Institute of Medicine, 31, 79, 94, 112
IPAV. *See* Informed Parents Against Vaccine-Associated Polio (IPAV)
Ireland, measles outbreaks, 92

Jacobson, Henning, 128–130
Jacobson v. Massachusetts, 127–130
Japan, pertussis outbreak, 16
Jauncey, Charles, 33
Jenner, Edward, 106–107, 112, 114, 117
Jenner or Christ? (pamphlet), 122
Johnson, Samuel, 37
Joseph P. Kennedy Foundation, 136–137

Kalus, Harry, 2
Karolinska Institute, 29
Karoly, John, 91
Karoly, Peter, 91
Katz, Sam, 114
Kendrick, Pearl, 25, 26
Kennedy, Edward, 136, 137
Kennedy, John F., 136, 184
Kennedy, Robert F., Jr., 152
Kennedy, Rosemary, 136–137
Kerensky, Michael, 124
Kerridge, David, 15
King, Ashley, 195
King, John and Catherine, 195
King, Larry, 123, 154, 160, 161, 163–164, 170
Kinnear, Johnnie, 33–35
Kinsbourne, Marcel, 100
Kirkman Laboratories, 153
Kluger, Jeffery, 152
Koch, Robert, 117
Koplan, Jeff, 72
Koppel, Ted, 113–114
Krigsman, Arthur, 101, 102
Kupsh, Debra, 141

Laitner, Jeanne, 194
The Lancet (journal), 95, 96
Langman, Rod, 174
Larry King Live (television program), 123, 154, 160, 162, 163–164
Lawsuits associated with DPT vaccine, 18–23, 32–40. *See also* Vaccine Injury Compensation Program (VICP)
Lawyers, anti-vaccine movement and, 124, 154

Lederle Laboratories, 19, 20–21
Leicester anti-vaccine rally, 111–112
Levine, Seymour, 35
Levy, Louis, 142–143
Levy, Sandra, 142
Lillard, Harvey, 118
Lister Pharmaceuticals, 33
Little, Kenneth, 134
Little, Lora, 134–135, 136, 180, 183, 202
London School of Hygiene and Tropical
    Medicine, 17
Los Angeles, measles outbreak, 138, 139
Louisiana, measles outbreak, xvi
Loveday, Susan, 35–40
*Loveday v. Renton and Wellcome
    Foundation Ltd.,* 35–40

Machin, Anthony, 35
Mad-cow disease, 182–183
Madsen, Kreesten, 94
Madsen, Thorvald, 14
Magid, Laurie, 91
Maher, Bill, 165–168, 202
Maier, William, 142
Maine, Hib meningitis outbreak, xii
Male Practice (Mendelsohn), 48
Mallon, Mary, 130–133
Mandatory vaccination, 139
    with influenza vaccine for healthcare
        workers, 196–197
    See also Compulsory vaccination
March of Dimes, 55, 162, 184
Margiotta, Joseph, 141
Marketing strategy, anti-vaccine
    movement and, 124–125
Maryland, repeal of religious exemption
    health laws, 195
Mason, James O., 19–20
Massachusetts
    repeal of religious exemption health
        laws, 195
Massachusetts Compulsory Anti-
    Vaccination League, 134
Mass marketing, anti-vaccine activists
    and, 122–123
McCarthy, Jenny, 149–154, 202, 214
    appearance on *The Doctors*, 155,
        156–157
    Barbara Loe Fisher's activism
        compared to, 151–154, 162

Carrey and, 163–164, 168
    "Green Our Vaccines" rally, 116–117,
        172, 179
    on vaccine ingredients, 182
McCartney, Thomas, 140
McCauley, Charles, 133–134
McConaughey, Matthew, 163
McCormick, Marie, 112
McDowell, Allen, 51–53
McKenzie, John, 69–70
McKown, Ian, 195
McKown, Kathleen, 195
Measles, xiv–xv, 152
    immunization rate and herd immunity,
        xvii
    incidence among unvaccinated
        children, 143
    outbreaks, xiv, xv–xvi, 92, 138, 139,
        185, 197–198, 207
    program to eliminate from United
        States, 136–138
Measles-mumps-rubella (MMR) vaccine,
    26
    autism and, 92–94, 97–98, 99,
        150–151
    early acceptance of, 191–192
    ingredients, 183
    Sears' alternative schedule and, 175
    vaccine court rulings, 87
Measles vaccine, decline in manufacturers
    of, 21
Measure for Measure (Shakespeare), 81
Medical advances, anti-vaccine movement
    and fear of, 120–121
Medical Research Council, 29
Meechan, Robert, 6
Mendelsohn, Robert, 5, 12, 47, 48–49
Mental retardation
    causes of, 40–44
    pertussis vaccine and risk of, 28–32
Merck, 73, 74, 183, 201, 205, 206
Mercury. *See* Thimerosal
Mesmerism, 117
Messonier, Nancy, 199
Metchnikoff, Elie, 120
Mica, Dan, 9
Mica, John, 9
Michigan, measles outbreak, xvi
Microbe Hunters, 167
Middlehurst, Donna, 7

Miller, David
    criticism of study, 28–32, 38–39
    study on pertussis vaccine-brain
        damage link, 17–18, 28, 36
    as expert witness, 33, 34
Miller, Zachary, 211
Minnesota, Hib meningitis outbreak, xi
Minshew, Nancy, 158
Missionary Temple at Large of the
    Universal Religious Brotherhood,
    Inc., 142
Mississippi, pertussis outbreak, xiii
Missouri, measles outbreak, xvi
MMR. *See* Measles-mumps-rubella
    (MMR) vaccine
Moll, Frederic, 14, 34
Moms Against Mercury, 96
Monosodium glutamate, 178
Montagnier, Luc, 51
Morbidity and Mortality Weekly Report
    (CDC), xiii–xiv
Morris, Jean, 29–30
Mortimer, Edward, 10–11, 18, 49–50
MSNBC, 11
Multiple sclerosis
    hepatitis B vaccine and, 65–66, 68, 88
    vaccines and, 86–87
Mumps
    complications, xvi, xviii
    immunization rate and herd immunity,
        xvii
    outbreak, xvi–xvii
    See also Measles-mumps-rubella
        (MMR) vaccine
Murphy, Jerome, 5, 12
Murphy, Trudy, 96
Murrow, Edward R., 168
Myelination of the Brain in the Newborn
    (Rorke-Adams), 91

Nathanson, Neil, 96
National Anti-Vaccination League, 118
National Childhood Vaccine Injury Act,
    21–23, 83, 104
National Immunization Program, 112,
    199
National Institutes of Health, 158, 168
National Public Radio, 159, 207
National Vaccine Compensation Fund, 124
National Vaccine Information Center, 8,
    11–12, 53, 67, 124, 149

"The Natural History of an Immunization
    Program" (graph), 191–192
Nature (journal), 119
NBC Nightly News (television program),
    xi, 11
Nelmes, Sarah, 106
Neuzil, Kathy, 205
New England Journal of Medicine
    (journal), 88, 94, 141, 165
New Mexico, measles outbreak, xvi
New York
    Hib meningitis outbreak, xi–xii
    smallpox outbreak, 133
New York Times (newspaper), 1, 92, 133,
    134, 154, 159, 204
Nigeria, polio vaccination programs, xvii
Nightline (television program), 113–114

Octoxynol, 178
Oklahoma, Hib meningitis outbreak, xii
Omer, Saad, 143
Omnibus Autism Proceeding, 85, 97–104
Open Your Eyes Wide! (Higgins), 134
Oprah (television program), 123, 150,
    162, 171
Oral polio vaccine, 26, 78–79
Oregon, pertussis outbreak, xiii
Orenstein, Walter, xvii–xviii, 112, 139,
    141, 196, 199, 200, 203, 211
Orient, Jane, 113–114
Osgood, Charles, 74
Osler, John, 39–40
Oz, Lisa, 188, 189
Oz, Mehmet, 188–190

Palmer, Bartlett Joshua, 118–119
Palmer, Daniel D., 118
Paranoia, anti-vaccine movement and,
    112–114
Parikh, Rahul, 123
Parke-Davis, 55
Pediatric Red Book (AAP), 5–6, 47
Peet, Amanda, 158, 162
Pennsylvania
    Hib meningitis outbreak, xi
    measles outbreak, xvi
Pertussis
    immunization rate and herd immunity,
        xvii
    incidence among unvaccinated
        children, 143, 144

mortality rates, 17
outbreaks, xii–xiv, 16
severity of, 46–47
Pertussis vaccine
allegations of brain
damage/neurological illness from,
2–11, 13–16, 28–40
British government study of risk of,
17–18
creation of, 25, 26
decline in manufacturers of, 19–21
early acceptance of, 191
improvement in safety, 28
infantile spasms and, 46
lawsuits associated with, 18–23,
32–40
risk of epilepsy or retardation from,
28–32, 40–44
shortages of, 19–21
side effects, 26–28
SIDS and, 46, 53
Peterson, Shannon, 182
Pfeiffer, Immanuel, 128
Pfizer, 202
Pharmaceutical companies
anti-vaccine movement and, 151–152,
153, 183–184, 202–203
decline in number of vaccine
manufacturers, 19–21
vaccine lawsuits and, 19–21
2-Phenoxyethanol, 178
Philosophical exemptions to vaccination
requirements, 142–143, 196
impact of, 143
Phipps, James, 106–107
Physicians' Desk Reference, 45–46
Pickering, James, 128–129
Pickering, Larry, 199
Pisani, Amy, 200
Pitcairn, John, 134
Pitman-Moore, 55
"Pitt Expert Goes Public to Counter
Fallacy on Autism," 158
Pittsburgh Post-Gazette (newspaper), 158
Pneumococcal infections, 181–182
Pneumococcal vaccine, 68–71, 181
Pneumocystis carinii, 50
Polio
historical mortality rate, xviii
outbreaks, xvii, 141, 152, 184
risk of infection, 186

"The Polio Crusade" (television
program), 184
Polio vaccines, 167
decline in manufacturers of, 21
early acceptance of, 191
Nigeria and distrust of, xvii
oral, 26, 78–79
Oz on, 189
polio caused by, 55, 58–59
promotion of, 162
Pollack, T. M., 29–30
Pontifical Academy for Life, 122
Population immunity. See Herd
(population) immunity
Priest, Julian, 34
Prince, Sarah, 139
Prince v. Massachusetts, 139–140
Principia College, 141
Prions, 182–183
Protests, anti-vaccine, 110, 111–112,
116–117
Public Health Laboratory, 29
Public health officials, anti-vaccine
movement and, 110–111,
112–113

Quinn, Aidan, 163

Raggedy Ann doll, 135–136
Rallies, anti-vaccine, 110, 111–112,
116–117
Reagan, Ronald, 21, 184
Recombinant DNA technology, HPV
vaccine and, 120–121
Reiki, 188
Religious exemptions to vaccination
requirements, 139–142,
192–196
impact of, 143
Religulous (film), 166
Renton, George, 35
Resciniti, Anthony (Tony), 5, 9
Resciniti, Leo, 5, 9
Rescue Angels, 154
Rheumatoid arthritis, hepatitis B vaccine
and, 87
Richmond, Julius, 192
Roberts, Robin, 52
Roizen, Michael, 188
Rome, Leonard, 6
Rorke-Adams, Lucy, 101

Rotavirus vaccine
  intussusception and, 71–73
  Oz on, 189–190
  testing, 181
  vaccine program, 204–206
Royal College of Physicians, 14
Royal Society of Medicine, 14
Rubella, historic mortality rate, xviii
Rubella vaccine, xvii, 87. *See also*
    Measles-mumps-rubella (MMR)
    vaccine
Rubert, William, 193
"Ruining It for the Rest of Us" (radio
    program), 207–209
Russell, Keri, 162
Rutledge, Wiley B., 139–140

Sabin, Albert, 58–59
SafeMinds, 96
Salamone, David, 78
Salamone, John, 77–80, 81, 82, 199–200
Salamone, Kathy, 78, 80
Salisbury, David, 16, 39, 92
Salk, Jonas, 55, 58–59, 167
Salmon, Daniel, 143
Salmonella typhi, 131
San Diego, measles outbreaks, 207
San Francisco Chronicle (newspaper), 57
Sanofi Aventis, 201
Sawyer, William, 96
"Scared Stiff: Worry in America"
    (television program), 52–53
School immunization requirements,
    136–139
  exemptions, 139–143, 192–196
Schuchat, Anne, 199
Schwartz, Jeff, 7, 8, 9, 26
Schwartz, Julie, 7
Science and Health (Eddy), 141
Sears, Jim, 171
Sears, Martha, 171
Sears, Robert, 202, 208
  alternative vaccination schedule,
    171–176, 187, 188
  on aluminum-containing vaccines,
    176–178
  anti-vaccine themes, 179–184
  on chemicals in vaccines, 178–179
  on herd immunity, 184–185
  recommended schedule of doctor visits,
    177–178
  on risks of vaccine-preventable
    diseases, 185–186
Sears, William, 171
The Sears Parenting Library, 171
Seizures
  DTP vaccine and, 28–32
  fever and, 40
  See also Epilepsy
Sexton, Nicky, 66
Shakespeare, William, 81
Shane, John, 90–91, 98
Shelley, Mary, 120
Sheridan, Dorothy, 194
Sheridan, Lisa, 194
Sherr, Alan, 142
Shingles, 164
Shock, pertussis vaccine and, 45
Shorvon, Simon, 44
A Shot in the Dark (Fisher), 57, 61, 81,
    91, 180, 183
Shriver, Eunice Kennedy, 137
SIDS. *See* Sudden Infant Death Syndrome
    (SIDS)
Simmons, Betty, 139
Slate.com, 146
Smallpox, 105–106
  outbreaks, 127–128, 133–134
Smallpox vaccine
  anti-vaccine movement and, 105,
    109–111
  compulsory vaccination and, 108–109
  creation of, 106–107
  objections to in United States,
    127–130, 134–136
Snyder, Colten, 97, 98–99, 101–102, 103
Snyder, Kathryn and Joseph, 102
Snyderman, Nancy, 159
Social good *vs.* individual freedom,
    compulsory vaccination and, 130
Socioeconomic class, resistance to
    vaccines and, 123–124
Sodium deoxycholate, 178
Soper, George, 131–132, 133
Special masters, 85–86
Spencer, Edwin, 128
SSPE (subacute sclerosing
    panencephalitis), xiv–xv
Steiner, Rudolf, xiii
Stephenson, John, 33
Stewart, Gordon
  on cause of HIV/AIDS, 50–51

on pertussis vaccine, 5, 12, 15, 25, 34–35, 36, 38
on severity of pertussis, 46
Stork, Travis, 155–158
Stossel, John, 52–53
Ström, Justus, 14, 34
Stuart-Smith, Murray, 34, 35, 36–40
Sudden Infant Death Syndrome (SIDS), 37
  court cases on vaccines and, 86
  hepatitis B vaccine and, 67–68
  pertussis deaths and, 17
  pertussis vaccine and, 46, 53
Sunday Morning (television program), 74
Sunday Times (newspaper), 17, 93
Super Nu-Thera, 153
Supreme Court, *Jacobson v. Massachusetts,* 127, 129–130
Sutherland, Cora, 193
Swan, Matthew, 194
Swan, Rita and Douglas, 194
Swine flu epidemic, 76, 165, 167

Tatel, Stephanie, 146–147
Tayloe, David, 157, 200
Tetanus, 145, 185–186
Tetanus vaccine, 26, 191
Texarkana, measles outbreak, 138
Thimerosal, 96, 116, 169
  Omnibus Autism Proceeding and, 97–98, 99, 100, 102
This Week (television program), 15
Thompson, George, 131
Thompson, Lea
  career after *Vaccine Roulette,* 11–12
  honor from National Vaccine Information Center, 53
  Vaccine Roulette and, 2–7, 9, 18, 25, 27, 40, 45–54, 103
Thoughtful House, 94, 96, 101
ThreeLac, 153
Through the Looking Glass (Carroll), 99
Time (magazine), 1, 152
Titicut Follies (film), 1–2
The Today Show (television program), 6, 11, 159, 171
Tom, Melanie, 19
Toner, Kevin, 20–21
Toxoid, 26
"The Tragedy of the Commons" (Hardin), 144–145
Tuberculosis, 193

Tuberculosis vaccine, 55–56
20/20 (television program), 65–67, 68
Twitchell, David and Ginger, 195
Twitchell, Robyn, 195
Typhoid fever, 131–133
Typhoid Mary, 130–133

United Kingdom, measles outbreaks, 92. *See also* England
United States Court of Claims, 86
Unvaccinated children, effect on community, 145–147, 184–185, 207–214
U.S. News and World Report (magazine), 170
USA Today (newspaper), 92, 154
Usui, Mikao, 188

Vaccination. *See* Compulsory vaccination; Immunization; Mandatory vaccination
"Vaccination: A Conversation Worth Having" (Maher), 165–166
Vaccination rates. *See* Immunization rates
Vaccinations and Lockjaw (Higgins), 134
Vaccination schedules, 161–162
  alternative, 171–176, 187, 188
Vaccination Vampire (pamphlet), 109
Vaccine Adverse Events Reporting System (VAERS), 23, 180
  rotavirus vaccine withdrawal and, 72–73
Vaccine advocacy/advocates, 212
  anti-vaccine groups and, 200–202
The Vaccine Book (Sears), 171, 180–184, 185–187
Vaccine Injury Compensation Program (VICP), 21, 31, 53, 54, 154
  conspiracy theorists and, 204
  expert witnesses, 88–91
  Omnibus Autism Proceeding, 97–104
  unusual rulings, 86–88
Vaccine inspectors, 110–111
Vaccine mandates, campaign against, 104
Vaccine manufacturers
  decline in number of, 19–21
  market stabilization and, 104
Vaccine policy, pediatrician's, 198–199
Vaccine-preventable diseases, severity of, 46–47, 181–182. *See also individual diseases*

Vaccines
  aluminum in, 176–178
  autism and (*see* Autism)
  chemicals in, 178–179
  claims of phantom ingredients in, 182–183
  claim that are unnatural, 116–117
  concomitant-use studies, 187
  conspiracy theory and, 51–52, 203–204
  false claims of harm, 114–116
  fear of, 120–121, 192
  federal funding for, 137
  history of problems with, 54–56
  immunological challenge of, 26, 173–174
  mercury in, 96–98, 99, 100, 102, 169
  multiple sclerosis and, 65–66, 68, 86–87, 88
  religious beliefs and, 121–122, 192–196
  school immunization requirements, 136–139
  shortages of, 19–21
  side effects/allergic reactions, 57–59, 180–181
  social class and resistance to, 123–124
  testing, 181
  See also individual vaccines
Vaccine safety advocacy, 96
  Fisher and, 58–59, 80–83
  Salamone and, 77–80
Vaccine Safety DataLink, 181
Vaccine toxins, autism and, 152
VAERS. *See* Vaccine Adverse Events Reporting System (VAERS)
Varicella. *See under* Chickenpox
Vashon Island (Washington), xii, 210–211
Vermont, pertussis outbreak, xiii
VICP. *See* Vaccine Injury Compensation Program (VICP)
Vinnedge, Debi, 122
Virchow, Rudolf, 89
Virginia, measles outbreak, xvi
Virtue, Doreen, 150
Vitamin D, 189
von Recklinghausen, Frederich Daniel, 89
Voodoo Histories (Aaronovitch), 203

Wakefield, Andrew, 92–94, 101
  criticism of, 94–96
  photos, 93, 95
Walker, Laurie, 194–195
Walker, Shauntay, 194
Wallace, Amy, 159–160
Wall Street Journal (newspaper), 92
Walters, Barbara, 67
Waltner, James, 6
Warren, Charles, 131
Washington, measles outbreak, xvi, 210–211
Washington Post (newspaper), 80
Waxman, Henry, 22
Wellcome Foundation, 33, 35, 36
Wenk, Eugene, 35
Werderitsch, Dorothy, 88, 98
Werne, Jacob, 14
Wexler, Deborah, 200
White, Tyler, 18–19
Whitestone, Heather, 62–63
Whitney, Aubrey, 193
Whitney, Edward, 193
Whooping cough. *See under* Pertussis
Willard Parker Hospital, 132
Williams, George, 129
Williams, Kathi, 6–7, 8, 9, 10, 23, 26, 27
Williams, Lisa, 157
Williams, Ted, 63
Wilson, John, 13–16, 34, 37–38
Winfrey, Oprah, 170
Wired (magazine), 159
Wisconsin, measles outbreak, xvi
Wiseman, Frederick, 1–2
World News Tonight (television program), 61, 62, 64, 69–70, 168
Wright family, religious exemption to vaccination requirements and, 140
Wuerger, Heidi, 212–213
www.kartnerhealth.com, 153
Wyeth Laboratories, 19, 55, 200, 201

Yankovich, Abra, 4–5
Yarkin, Celina, 210–212
Yellow fever vaccine, 54
YOU: Having a Baby (Oz & Roizen), 188–190
Young, Bobby, 49

Zucht, Rosalyn, 130